DINOSAURS UNDER THE AURORA

Life of the Past

James O. Farlow, editor

DINOSAURS
UNDER THE AURORA

Roland A. Gangloff

Indiana University Press
Bloomington and Indianapolis

This book is a publication of

Indiana University Press
601 North Morton Street
Bloomington, Indiana 47404-3797 USA

iupress.indiana.edu

Telephone orders 800-842-6796
Fax orders 812-855-7931

♾ The paper used in this publication meets the minimum requirements of the American National Standard for Information Sciences—Permanence of Paper for Printed Library Materials, ANSI Z39.48-1992.

Manufactured in the United States of America

Library of Congress Cataloging-in-Publication Data

Gangloff, Roland A.
Dinosaurs under the aurora / Roland A. Gangloff.
p. cm. — (Life of the past)
Includes bibliographical references and index.
ISBN 978-0-253-00080-4 (cloth : alk. paper) — ISBN 978-0-253-00718-6 (ebook) 1. Dinosaurs—Alaska. 2. Paleontology—Cretaceous. 3. Geology, Stratigraphic—Cretaceous. 4. Geology—Alaska. I. Title.
QE861.8.A4G36 2012
567.909798'7—dc23

2012005738

1 2 3 4 5 16 15 14 13 12

CONTENTS

PREFACE

This book is written not just for the dinosaur enthusiast but for those readers that have an interest in the Arctic and Alaska. The discovery of dinosaurs in the Arctic of Alaska and the subsequent accumulation of a surprisingly rich record of these fossil animals challenged many widely held misconceptions about the ecology, biology, and biogeography of these fascinating beasts. These high-latitude discoveries also called into question the simplistic extinction scenarios that were advanced during the 1960s and '70s and that continue to fuel debate today. The Arctic of Alaska presented the author with unusual and exhilarating challenges. Some of the greatest difficulties stemmed from the size and remoteness of Alaska. However, the demands of working within the idiosyncratic world of Alaskan politics and economics make research of any kind in Alaska a truly unique experience. The reader will not only be treated to the excitement and exigencies that accompany paleontological field research in the Arctic environment but will also gain an understanding of just how the scientific process and scientists really work. In addition, the reader will get a feel and taste for Alaska, the place.

The first part of this story summarizes the last fifty years of the Arctic discovery of dinosaurs, whose remains are scattered throughout the vast circumarctic region. Then it goes on to describe Alaska's dinosaur record and to tell the fascinating story of their almost nondiscovery and its impact on the dinosaur extinction debate. Next, the narrative takes readers to the North Slope, introducing them to the methods and related challenges of fieldwork in Alaska's Arctic and detailing the extraordinary collection of dinosaur remains that has been accrued over the last twenty-five years. After placing these fossils and their interpretation in proper geologic, biologic, and time perspective, the book takes up the broader impacts and possible future directions of research on dinosaurs in Arctic Alaska and the rest of the circumarctic, considering how new technologies, global climate change, and the "cold war" that is erupting over the natural resources that have been sequestered beneath Arctic ice for eons may impact future research.

It is hoped that the reader will gain a new perspective on Alaska, the Arctic, and the scientific process and come away with a sense of how much has been learned about dinosaurs and their polar world some seventy million years ago.

ACKNOWLEDGMENTS

To the hundreds of volunteers, teachers, and students that gave their time, sweat, and blood to unearth the story of Alaska's Arctic dinosaurs, I am eternally thankful. Over my eighteen years of work in Alaska, I benefited greatly from the help and counsel of a host of colleagues. Some of the contributions of a number of these colleagues are described in the book, but I must acknowledge here the enormous contributions of time and valuable counsel of my Alaskan colleagues: Kevin May, Dave Norton, Anne Pasch, Don Triplehorn, Gary Selinger, and Gary Grassi. I must extend a special thank you to Canadian colleagues Philip Currie, Richard McCrea, David Eberth, Donald Brinkman, Darren Tanke, Grant Zazula, John, Storer, and David Evans, who, along with Thomas Rich, Patricia Vickers-Rich, Pascal Godefroit, Kyle Davies, M. K. Brett-Surman, Tony Fiorillo, and Mark Goodwin, generously shared their expertise, publications, and data. Jørn Hurum, John Tarduno, and Patrick Druckenmiller shared their valuable insights and field images. David Smith, Karen Carr, Buff and Gerald Corsi provided graphics and their valuable time. Judy Scotchmoor, Robert Sloan, and James Farlow provided valuable editorial assistance to help make this book better.

DINOSAURS UNDER THE AURORA

THE ARCTIC SETTING

1

The Arctic Coastal Plain circa Seventy Million Years Ago

The Arctic coastal plain is crisscrossed by a host of meandering rivers that drain the northern slope of the rugged ancestral Brooks Range to the south. The rivers are pregnant with organic-rich sediment and rush headlong to the northern sea, being fed by melting snowfields and the common cloudbursts that sweep in from the Western Interior Seaway to the north and east. Large herds or aggregates of duck-billed dinosaurs move along river banks feeding on dense "gardens" of mud-loving horsetail rushes that have sprung forth from their subterranean rhizomes into the reawakening sunlight.[1] Monodominant patches of drought-resistant ferns are interspersed with clumps of herbaceous angiosperms and grasses on small ridges and levee slopes.[2] Stacks and tangles of lichen and moss-encrusted logs and branches form along the edges of sloughs and oxbow lakes. Large logs of deciduous conifers, such as *Parataxodium* from forests deep in the interior, have mixed with smaller ones that had spent their lives closer to where they now lie.

The rushes and ferns grow very rapidly due to the long, warm, lengthening days. The largest concentrations of duckbills are located along the margins of shallow sloughs, floodplain lakes, ephemeral ponds, and river banks with smaller groups strung out along the crowns of larger river levees. The rich bounties of aquatic plants that occupy the floodplain lakes and ponds are the targets of groups of duckbills like *Edmontosaurus*, who can use their spoonbill-like jaws to "shovel" in rhizome mats and clusters of plants similar to water ferns, duckweed, and water lilies.

Some small groups of duckbills venture into upland coniferous forests and come upon larger groups of ceratopsians such as *Pachyrhinosaurus*. The ceratopsians and duckbills selectively partake of mixes of deciduous conifer trees such as *Parataxodium* and *Podozamites*. Some hadrosaurs and ceratopsians feed on the leaves of these trees as well as on an understory of ferns, scattered cycadophytes (*Nilssoniocladus*), and the rarer clusters of mistletoe and sandalwood. Individual ceratopsians such as *Pachyrhinosaurus* select fruits such as *Ampelopsis* and *Cissites*.[3] On occasion, they take in insects and snails crawling over the leaves—a serendipitous protein treat. Only adults and subadults venture deep into the denser wooded areas and upland forests where *Troodon* and other gracile predators can hide and wait in ambush. These woodlands are too dangerous for the more vulnerable juveniles to feed in.

Hours later, the sky turns to an angry gray, filling with roiling masses of water-laden clouds that are being driven southward against a dark wall of mountains. Torrents of rain sweep down the mountain slopes and are channeled northward to the sea. A 100 miles away, the groups of *Edmontosaurus* are totally absorbed by their search for food, unaware of the rush of water that is coming their way.

A large group of *Edmontosaurus*, a few standing conspicuously above the rest, appears at the top of a large levee and stops to feed on the lush plant smorgasbord before descending to the edge of the fast-flowing river. While their fellow travelers are dining, a dozen of the larger individuals approach the river. Hesitating for just a moment, they plunge in, making enormous splashes as they push against the strong current. The rest of the group is now dominated by juveniles that are half to one-quarter the size of their adult and subadult attendants. Just as the adults finally reach the opposite shore and struggle up the opposite embankment, a larger group of adults and subadults catch up. Anxious to follow their leaders, they take the plunge, and this spurs the juveniles to follow. The water is shallow enough that most of the juveniles can just touch bottom. Just as this mix of adults, subadults, and juveniles makes it a quarter of the way across the river, a four-foot-high pulse of water reaches the struggling group, greatly increasing the strength and velocity of the current. Most of the larger animals struggle across, making ever-widening arcs, and end up hundreds of feet farther downstream than those who first crossed. The juveniles, who are only one year old or younger, are quickly exhausted by the struggle against the current and are unable to touch bottom. The first youngsters flail about wildly and begin to drown as they are carried down river. The remainder of the nursery that is onshore continues to plunge into the tumultuous waters, as panic takes over and instinct drives them to stay with the rest of their kind.

Days later and miles farther downriver, hundreds of bloated bodies are strewn along the main river banks and in sloughs. Large rafts of bloated and rotting bodies form in sloughs that are partially isolated by the lower water.[4] However, the greatest mass of "bloat and float" has accumulated in a large shallow lake that formed when a part of the levee near the crossing point gave way at the height of the flood. As the levee wall gave way, the water rushed through the gap and pulled many of the juvenile and a few subadult bodies with it. The newly formed shallow lake became clogged with hundreds of rotting corpses. The stench of so many rotting carcasses draws a host of opportunistic theropods, such as *Troodon*, *Dromaeosaurus*, and the larger *Albertosaurus*, who are eager to take advantage of the available feast. It is easy for these meat eaters to tear away masses of softened muscle, and they seldom penetrate to the bone in the process. There are so many bodies that the scavengers have little need to eat down to the bone on any one body. As the floating carcasses decay further, partial limbs, sections of tails, and sometimes whole sections of their bodies fall away and are quickly buried. A volcanic eruption hundreds of miles distant has helped to seal the nursery group's sedimentary "coffin," the volcanic debris that has rained down from the sky having raised the sediment load carried by the river.

Fig. 1.1. Cartoon of juveniles being separated from adults and subadults and then swept away by current during a pulse of high water during early spring thaw. *Credit: Tom Stewart and the University of Alaska Museum of the North, Fairbanks.*

How do we know that such a description may have been an ancient reality? What is based on hard data, and what is logical and reasonable extrapolation of these data? Paleontologists try to reconstruct life on this Earth as it existed in the ancient past—not by constructing some static skeletal exhibit but rather by portraying all of the dynamic interactions within the environment that nurture and shape physical existence. It is their passionate desire to assemble all of the fossil evidence and, in the context of the physical and chemical geologic evidence, journey back in time. Modern paleontologists also use an acquired knowledge of the world as it can be observed to function today to infer the past.

The purpose of this book is threefold. The first is to summarize the record of Arctic dinosaurs that has accrued over the last half century in concert with a review of the evidence that underlies that record. The second is to describe the methodology and challenges of dinosaur field research in Arctic Alaska, as well as interpret the evidence of dinosaurs that has been put together for Arctic Alaska and closely related high-latitude areas over the last twenty-five years. The third is to discuss the possible directions that Arctic dinosaur research will take in the future. However, these goals cannot be properly accomplished without an introduction to the Arctic as a setting or context. Because the Arctic and Alaska are still relatively poorly known to most dwellers of the lower latitudes, it is important that the reader be given a better understanding of the environment in which these records of dinosaurs have been compiled. In addition, the body of mythology and misrepresentation that preceded this record of dinosaurs and has subsequently attended the discoveries and scientific analysis of these fascinating beasts should not be ignored. Now, let's first experience the Arctic as a place, a milieu.

The Arctic: Realm of Myth and Mystery

The Arctic, in winter, is a place that seems to be perpetual night accented by kaleidoscopic colors that dance across the starry sky unconstrained by static geometry or human imagination. At times, it appears to be wholly comprised of nearly flat snowbound surfaces reaching from horizon to horizon. It is home to a sun that in summer never sets below the Earth's edge but rises in a spiral and then slips below that surface for the other half of the year. It is an immense region, where the land surface and circumferential ocean are dominated by ice and snow for eight to twelve months each year, a venue that seemingly swallows up most who dare to traverse it. A place where strange ice-entombed beasts abound, whose countless bones appear to be scattered about as the result of great upheavals of earth and sea. Immanuel Velikovsky, in his book *Earth in Upheaval*, cites the abundance of fossil animals such as mammoths in Alaska and Siberia as evidence that great cataclysms like giant sea waves must have taken place.[5] He goes on in the same chapter to incorrectly claim that mummified mammoth remains in Siberia contained plants that were not to be found in the Arctic and that the mammoths must have been thrust into the Arctic from lower latitudes by great earth upheavals. This idea of the Arctic prevails in the mind of the uninitiated; it is the Arctic of popular lore and literature that only partially or incorrectly portrays the nature of this vast and vital region. The Arctic—that part of the Earth that lies above the Arctic Circle at 66.5° N—is defined in more than one way (see plate 1).[6] Much of the Arctic, through most of human history, has been without permanent settlements or has been the venue of only a few seekers of game or fabled riches. It is no surprise that the Arctic has spawned many misconceptions and continues to be a place of mystery and wonder.

The Arctic has been a region of mystery to Europeans ever since their early forays into this forbidding land in the early eighteenth century. Prior to that, it was a region that only the dreaded Vikings, among Europeans, were able to partially tame and settle. Little was known of the waves of migrations and settlements that marked the history of Asian peoples in the circumarctic. The rise of worldwide commercialism, as a result of the Industrial Revolution in Europe, and the spread of Europeans into northern North America and Siberia in the eighteenth century gave birth to the search for a Northwest Passage from the Atlantic to the Pacific Ocean. This quest, based on little concrete geographic information and aptly titled the "voyages of delusion," resulted in over two hundred years of failed expeditions; ships and men disappeared, only adding to the mythology of the "frozen north."[7] Even the celebrated explorer Robert Peary fell victim to the common Arctic atmospheric high jinx of mirage and reported in 1899 the sighting of a new land "TrueJesup Land" west of Ellesmere—that was never seen again. It wasn't until 1969 that the ice-breaking tanker

Manhattan finally transited this long sought-after route on an experimental basis. The *Manhattan* plowed through the sea ice, winding its way along the southern edge of Canada's Arctic islands and the northern Canadian Arctic coast. However, the first commercially feasible navigation of the Northwest Passage still awaits further sea ice retreat that the present trend in global weather change will bring.

The reader at this point might be left with a somewhat negative view of this fascinating environment. The natural history of the Arctic is awe inspiring and borders on the mystical in the way that it stimulates the senses and excites the imagination. This combination of poetry and natural history is beautifully captured by Barry Lopez in his *Arctic Dreams*, an inspiring poetic essay that reflects Lopez's deep connection with the landscapes and animals of the Arctic.[8] Lopez instructs while taking the reader's breath away with his powers of description.

The Arctic is a uniquely stimulating and beautiful place. You are struck by the stark contrasts that abound during every yearly cycle, the myriad hues of white and blue of the Arctic ice-covered sea and land, and the array of textures that snow displays as well as the complex swirling dynamics that it can assume as it is blown across the whitened tundra. You can be totally mesmerized by the frenetic pulses of an aurora-streaked sky, whose curtains of color are constantly changing their tints and shapes. It is easy to miss the subtle reflection of auroral colors that highlight the snow around you as your gaze is drawn to the show above. In the Arctic summer, the seemingly eternal cloak of dark and twilight gives way to almost endless daylight that resets your circadian rhythms. A winter's white vastness turns into an ocean of variegated greens that then give way to a blazing palette covered with intense reds, browns, and yellows as an autumnal wave sweeps southward across the tundra. As a geologist who has worked in hot and warm deserts as well as the Arctic, I am fascinated by the great expanses and clarification of form and detail that rock and sediment surfaces assume in both of these environments. However, the transparency of the atmospheric conditions that you can encounter in the Arctic greatly enhances the experience. In both environments, the scale of what you are seeing is difficult to grasp.

In my work as a field paleontologist, I have found the wildlife of the Arctic summer to be a constant source of wonder and delight. The influx of birds from all over the globe during the summer, for example, is amazing to see. Birds in dizzying numbers and variety descend on the rivers, countless lakes, and bluffs in concentrations that can only be matched to the south along major flyways or refuges. Many a night can be accompanied by the soothing refrain of a loon. In Alaska and other parts of the Arctic, vast herds of deer such as caribou or reindeer can approach the numbers and exhibit the dynamics of herds on the plains of east Africa during migration time. When you spend enough time in the field, you can find yourself in the right place at the right time confronting a tight defense ring of shaggy-coated musk-ox with their strangely primitive-looking helmet and horns—a sight that connects us with our Pleistocene Ice Age ancestors. Sometimes you are treated to more intimate encounters, such as the antics of two Arctic fox pups playing "tag" around a clump of tall grass or an Arctic ground squirrel outside the supply tent about to abscond with the last package of ramen.

Alaska's Place in the Arctic

Alaska as a whole takes on much of the aforementioned cloak of mystery and wonder even though it is only partially enfolded by Arctic boundaries (see plate 1 and figure 1.2). A long history of commercial exploitation by Europeans was brought about by gold rushes, land rushes, and, most recently, energy rushes. Alaska is a place that feeds the dreams—dreams like those that fueled the great migrations to the American West in the 1800s—of treasure hunters, large and small. It is a place that seemingly offers total freedom if you only rely on your own skills and subsist on the bounty of the land. Alaska today and yesterday is more myth than fact and probably will remain so.

Every year thousands of tourists flock to Alaska's interior and Arctic coast during the cold dark winter to spend time beneath the "northern lights," hoping the exposure will bless them with long life and/or fertility. The magical properties assigned to the aurora borealis are only some of the modern myths that are ascribed to what tourist brochures refer to as the "last frontier."

As a whole, Alaska is still envisioned by many outsiders as the "land of perpetual ice and snow"—a land that is plunged into darkness by a fleeing sun for over half of the year. Never mind that the other side of the yearly coin produces the so-called midnight sun and that in reality, only the northern third of the state is subjected to a sun that sinks below the horizon for nearly half of the year (and even this is marked by twilight and indirect sun at each end of the "dark" period). Never mind that, like Arctic Siberia, Arctic Alaska can experience midsummer temperatures ranging from 80 to 90° F (26 to 32° C). The emphasis on ice and snow in visualizations of the Arctic and Alaska ignores the reality that the Arctic is a cold desert. Most of the central and northern Alaskan interiors are semiarid to arid in their annual precipitation counts. It is only in the high mountains and along the coasts that snowfall commonly reaches depths beyond a few feet, and much of that disappears by late June. Like the Arctic that enfolds its northern third, Alaska's image is distorted by perceptions that are born of lack of knowledge or a mysticism that is too often reinforced by movies and popular literature. Movies such as *The Thing from Another World* and *Superman* are only two examples.[9] Why, you may ask, is a book about Arctic and Alaskan dinosaurs recounting the myths associated with these areas? Because a modern set of myths and misconceptions had a role in the discovery and subsequent recognition of the first bona fide Arctic Alaskan dinosaurs. Myths and misconceptions pertaining to the state of preservation and the paleoecological implications of Alaska's dinosaurs have been published time and again on the World Wide Web and in popular books about Alaska. Examples range from James Michener's popular 1988 novel *Alaska* to a more recent book by three literal creationists as well as books by flood geologists.[10] In chapter 11 of *The Ice Castle,* Michener abandons his solid research and perpetuates the popular misconception that dinosaur

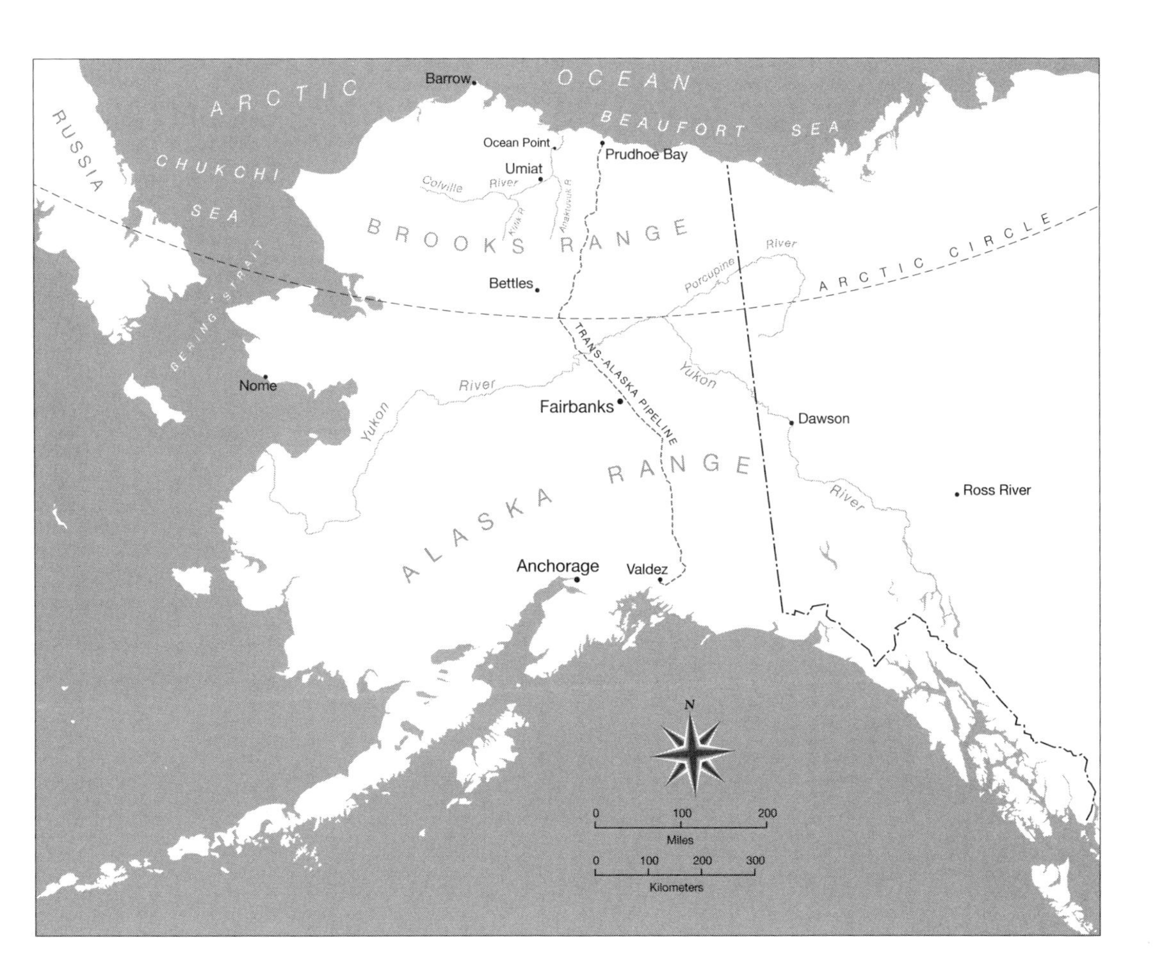

Fig. 1.2. Map of Alaska and adjacent parts of the Yukon Territory with pertinent geographic features and sites. *Credit: Dixon Jones.*

bones are usually petrified and contain only minerals and are devoid of original constituents. Michener also repeats a myth that is still found in creationist literature and on creationist websites to the effect that unpetrified or unmodified "real bone" of dinosaurs was found in northern Alaska. This is a myth that was started by misinterpretations or distortions of early news reports describing the first reports of dinosaurs from the North Slope of Alaska.

Mammoths and Dinosaurs: Separating Fact from Fallacy

In many people's minds, dinosaurs and extinct Pleistocene-age mammals, especially mammoths, are often confused with one another.[11] In order to be sure that the reader understands the difference between these two extinct animals, I introduce two working definitions. Dinosaurs are now considered to be a class or major group of vertebrate animals that are part of a major evolutionary lineage called the Archosauria. Dinosaurs are distant "cousins" of most living and fossil reptiles. Their closest living relatives are fossil and living birds and crocodilians. Birds are now considered a living extension of the dinosaur lineage. Dinosaurs are defined chronologically as having evolved from ancestors in the early part of the Mesozoic Era, some 220–230 million years ago. They greatly diversified throughout the era until most lineages died out at the end of it, around sixty-six to sixty-five million years ago. When the term "dinosaur" is used in this book, it refers to nonavian dinosaurs. The word "mammoth," as used in this book, refers to an extinct member of a group of mammals called proboscideans. Mammoths are closely related to living elephants and first appeared in Africa during the later part of the Cenozoic Era, three to four million years ago. The vast majority became extinct nine to ten thousand years ago at the end of the Pleistocene Epoch.

The woolly mammoth has become a popular icon that is symbolic of Alaska and much of Arctic and subarctic Siberia. As the curator of earth sciences at the University of Alaska Museum in Fairbanks, I was responsible for collections of fossil vertebrates such as mammoths and dinosaurs. In my dealings with both native and nonnative Alaskans, I often came across individuals within both groups who made no distinction between dinosaurs and fossil mammals such as mammoths. In the course of my researching the history and ethnography of the Arctic and the development of ideas regarding both dinosaurs and mammoths, it became clear to me that this confusion had its roots in cultural myths and misconceptions that were engendered by some of the earliest human encounters with these extinct prehistoric animals.

The discovery of intriguing and strange beasts in the Arctic first entered the consciousness of people in the lower latitudes through reports and drawings filed by tourists, merchants, and natural historians who were exploring the Arctic for economic and intellectual treasures. Mammoths, along with other large Pleistocene-age mammals, were being found in great numbers along the banks of Siberian rivers in the Arctic and subarctic of northeastern Eurasia. Reports, drawings, and some bones and teeth of mammoths first reached Europe in the late eighteenth century and at times led to rampant speculation that formed the basis of popular mythology.[12] As more and more evidence of mammoths and associated extinct mammals reached the cities of Europe, it stimulated intriguing questions such as, were such beasts still roaming the Arctic? Were these creatures left over from a lost world?[13] Could they be the poor wretches that had survived the great biblical deluge of Noah—corroboration for the Book of Genesis? Early writings and illustrations equated them with unicorns and other members of the European mythic bestiary.[14] By the late nineteenth and early twentieth centuries, the frozen and mummified remains of mammoths were reaching kingly curiosity cabinets and museum collections where scholars could study them. Eventually, word of such strange beasts spread throughout the populace, resulting in the word "mammoth" being incorporated in the popular lexicon, which came to be equated with anything huge and ultimately became synonymous with "behemoth."

Prior to their first contacts with Europeans, native peoples of Siberia and Alaska had come across large fossil bones piled like logjams along the banks of the great rivers or projecting from thawing permafrost-encapsulated sediments. As these remains appeared to be those of animals but different from any of the living animals that Arctic natives had hunted and studied, they often assigned them to a subterranean world, hypothesizing

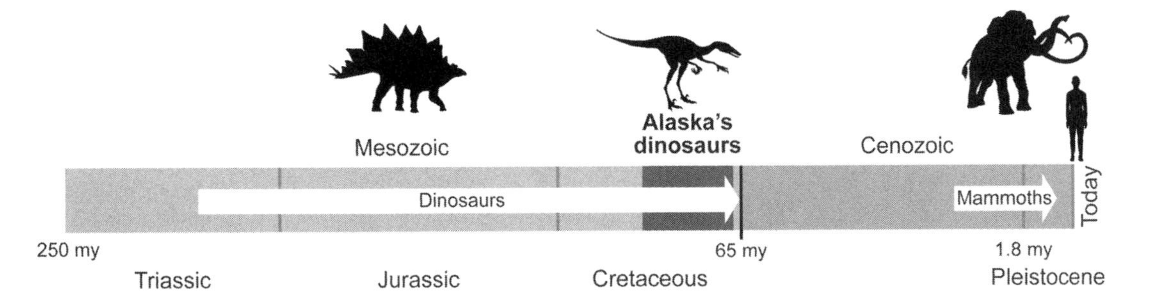

Fig. 1.3. Dinosaur-mammoth timeline with Alaska's North Slope dinosaur range in geologic time. *Credit: David Smith.*

that these creatures must have lived there like the more familiar rodents. These strange and often gigantic beasts became part of a native mystic realm, part of the shaman's kit for native peoples on both sides of the Bering Strait.[15] The ivory tusks of mammoths and their Pleistocene cousins, the mastodonts, became prized objects that were integrated into aboriginal artistic and religious traditions—a striking convergence with the European integration of mammoths and other Pleistocene mammals into a mythic bestiary and religious traditions such as the biblical behemoth.

The Science of Paleontology Emerges: Displacing Myth as the Fossil Record Grows

As Europeans and other foreigners spread throughout the Arctic from the late nineteenth to mid-twentieth centuries looking for gold, furs, and land, they developed an extensive trade in fossil ivory that was turned into piano keys and highly prized objects d'art. Eventually, more and more remains of extinct prehistoric animals, such as bones, teeth, and mummified skin, reached museums and the scrutiny of scientists, who studied them and began to recognize their relationships to living animals. This recognition of relationships evolved hand in hand with the discovery of previously unknown extant animals such as the elephant and rhinoceros as new lands were explored in both the Old and New World. This development of the early foundations of what would eventually be called science was known as natural philosophy and natural history. It was only a matter of time before the first dinosaurs would be found and described.

The discovery of the first remains of dinosaurs and the subsequent study and coining of their name took place during the period between 1822 and 1841.[16] These discoveries of ancient animals that were no longer part of the living fauna led to our understanding of important concepts such as extinction and to the recognition that fossils were the witnesses to the progression of life on the planet—a tenet of modern geology and paleontology.[17]

The science of paleontology grew out of the European intellectual revolution that spanned the seventeenth, eighteenth, and nineteenth centuries and was the result of the recognition that fossils were the direct evidence of once living organisms and that their enclosing sediments held many of the keys to understanding the world these organisms lived in. In consort with paleontology, geology developed its fundamental principles, such as superposition, original horizontality, and uniformitarianism. Geologists integrated the laws and principles of physics and chemistry to establish a sound and rational framework for interpreting the Earth's history and processes. By the beginning of the twentieth century, these efforts and the fossil data compiled by paleontologists had yielded a time scale that was immense and far reaching in its magnitude and implications. Paleontology matured significantly when it helped to define and then integrate the biological principles of organic evolution that emerged from the studies of Cuvier, Lamarck, Wallace, and Darwin and the developing geologic time scale. The development of the natural world could no longer be confined to a few thousand years or dominated by "demons," to paraphrase the late Carl Sagan.[18] However, as I pursued my collecting and study of the dinosaurs of Alaska and the Arctic, I realized that a modern body of myths still holds these fascinating animals (dinosaurs and mammoths) captive. This mythology and body of misconceptions/misrepresentations persist despite the amazing numbers of astounding discoveries and detailed studies of both that have been conducted ever since the late nineteenth century. Let us now take a tour of the dinosaur-producing sites in the Arctic, beginning in the eastern Arctic and ending in the western Arctic. Hang on, it is going to be an exciting, and hopefully enlightening, whirlwind tour.

TRACKS LEAD THE WAY

2

Circumarctic Dinosaur Discoveries from Svalbard to Koryakia

Geologists . . . inhabit scenes that no one ever saw, scenes of global sweep, gone and gone again, including seas, mountains, rivers, forests, and archipelagos of aching beauty, rising in volcanic violence to settle down quietly and forever disappear—almost disappear.

—JOHN MCPHEE, *Basin and Range*

The following account of the history of dinosaur discoveries in the Arctic begins in the Svalbard Archipelago in the eastern Arctic, where the first evidence of Arctic dinosaurs was found, and then proceeds to the western Arctic and sites in Canada's Arctic Islands, the Northwest Territories, and the Yukon Territory. From here discoveries in eastern and northeastern Siberia are detailed in an attempt to set the stage and develop a proper perspective for the more extensive treatment of the discovery of dinosaurs on Alaska's Arctic North Slope.

The Svalbard Archipelago

Dinosaur fossils were first discovered in the Arctic by scientists in a most unlikely archipelago of isolated islands now known by the Norwegian name Svalbard. Originally referred to in Dutch as "Spitzbergen," the name, which means "pointed mountains," refers to the largest island's extremely rugged topography. This barren and wind-swept group of islands are north of the Fenno-Scandinavian Peninsula, due west of the Russian islands of Novaya Zemlya and due east of Greenland (see plate 1).

Svalbard is near the western margin of the eastern Arctic and is one of the largest pieces of land in a land-poor section of the Arctic Ocean. Svalbard has, for most of its recent history, been just a few degrees south of the great Arctic ice cap. The present global warming trend has produced a dramatic thinning of the ice and a poleward retreat of the ice cap margin over the last decade.[1] Although the Svalbard Archipelago is presently more than 10° N of the Arctic Circle, it was at or below the paleo-Arctic circle during the Early Cretaceous.[2]

The year is 1960 and the twenty-first International Geological Congress is to be held in Norway. Prior to the meeting an international group of geologists and paleontologists have the rare opportunity, as part of an

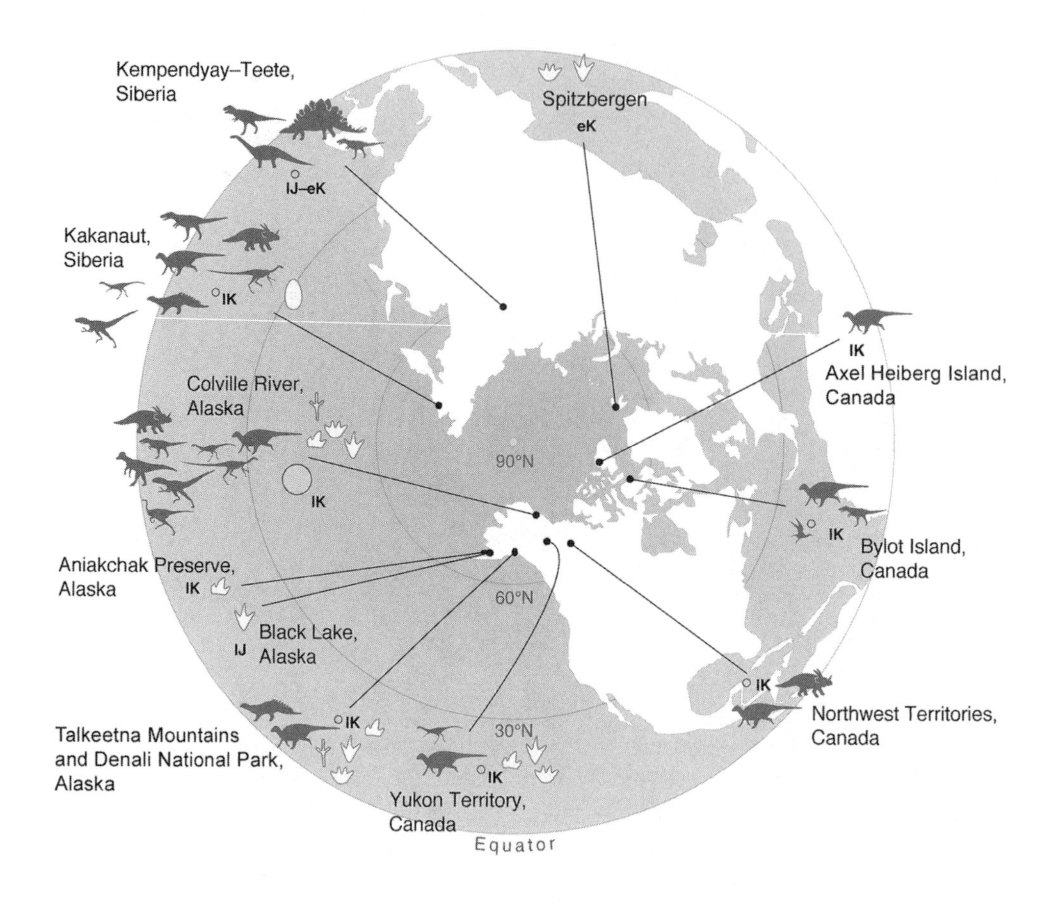

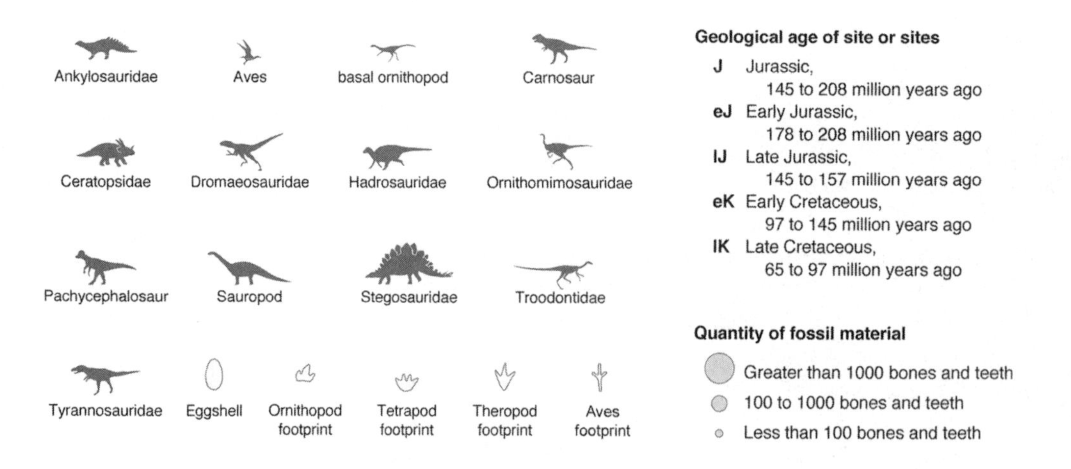

Fig. 2.1. North polar dinosaur distribution map listing types of dinosaurs (avian and nonavian) and ichnofossils. *Credit: Dixon Jones and David Smith (modified from Rich, Vickers-Rich, and Gangloff 2002).*

internationally sponsored field excursion, to look at exposures of rocks near the top of the world that span a huge chunk of geologic time. Most importantly, the rocks represent a critical time span that includes the evolution of the dinosaurs and their direct ancestors. Earlier geologic mapping and field studies on the largest island of Spitzbergen in the Svalbard Archipelago had established the presence of both marine and terrestrial rock units as well as their age. The trip is one of several that have been scheduled as part of the

Fig. 2.2. Spitzbergen-Svalbard. View across Sassenfjorden toward Billefjorden. Block of Festningen sandstone in foreground. *Photo credit: Patrick Drukenmiller.*

International Geological Congress meeting.[3] As is often the case in science, serendipity plays a significant role in the discovery of new evidence. The field party came to the rugged fjord-rich southwestern part of Spitzbergen by ship in August. The excursion vessel, the *Valkyrien,* acted as a dormitory and base for the participants who made daily trips to explore the spectacular outcroppings of rocks along the walls of the fjords. It was near the end of one of these day trips to the Grønfjorden that two of the scientists, Albert de Lapparent and Robert Laffitte, found themselves at the top of a steep slab of sandstone that plunged onto a narrow beach below. They carefully climbed down to the base of the sandstone slab in order to study the sequence of rocks from a less precarious position. As their eyes adjusted to the low oblique light and the surface texture of the sandstone, they were astonished to see a series of thirteen large three-toed footprints incised in the rock, part of a trackway.[4] The two scientists were surprised because they were supposed to be in marine rocks and because the field guides made no mention of fossil footprints. The two were quickly joined by others who heard their shouts. Not expecting to find fossil footprints, no one in the party had chalk to outline the ichnites so they could get clear photographs. However, placing white algal-covered pebbles around each imprint did the trick. Documentation was enhanced with sketches and measurements before the group returned to the *Valkyrie*—a ship whose name is quite appropriate for conveying those who seek the remains of the prehistoric dead, for it is the Valkyrie—handmaidens of the Norse god Odin—that chose and then escorted the heroic dead to Valhalla and everlasting glory. If the scientists had landed at a different spot, if the sandstone beds had been horizontal, or if it had been a different time of day, the footprints could have been missed. That the circumstances had to be just right may be why earlier field expeditions had overlooked these tracks. Subsequent field and laboratory studies of the same rock unit have established that there are several stratigraphic levels with very similar tracks and that they most probably represent a group of bipedal dinosaurs known as ornithopods.[5] The Ornithopoda include such taxonomic groups as the hadrosaurs (duckbills), iguanodontids, and hypsilophodontids. Hadrosaurs, iguanodontids, and hypsilophodontids form a group of closely related bipedal and semibipedal herbivorous dinosaurs that are distinguished by the development of a complex array of tooth and jaw mechanisms. These dynamic jaws allowed them to efficiently chew a range of tough vegetation such as ferns, horsetails, and cycads. Hadrosaurs also developed an extraordinary set of tooth batteries (see figure 6.5) that allowed for continuous tooth replacement. All of these ornithopods developed a predentary bone that was part of a new way to gather or "nip" vegetation that reached an acme in the duck-like toothless bill of hadrosaurs. All members of this lineage also developed a complex of ossified tendons that cross-braced their vertebral column and reflected the evolution of their bipedal and semibipedal ways of walking.

A later expedition, in 1976, that was focused on the study of sediments and sedimentary rocks at a locality some 75 miles (120 kilometers) away on the eastern side of Spitzbergen discovered two footprints.[6] These belonged to a bipedal dinosaur that likely was a large meat-eating theropod such as a *Megalosaurus*. A quick review of the pertinent literature reveals that these are, indeed, the first evidence of dinosaurs to be documented from the Arctic.

The Svalbard Archipelago experienced a wetter, warmer, and more equitable climate in the Early Cretaceous than it does today. Even being at or near the paleo-Arctic Circle, the area was covered with polar deciduous forests that were equivalent to boreal or cool subtropical forests of today. The presence of coals, megafossil plant, and animal remains provides evidence of such forests.[7]

The initial discovery encouraged efforts that led to the discovery of two major divisions of dinosaurs and several track sites contained within a thick stratigraphic interval that reinforce the hypothesis that dinosaurs were abundant, widespread, and thriving in the paleo-Arctic during much of the Cretaceous period. In addition, these findings have engendered several ongoing research programs that hold a solid promise of further discoveries that will enrich the history of dinosaurs throughout the Arctic.

Bylot Island

Just east of the Canadian Arctic Archipelago lies the obscure Bylot Island and the Canadian territory of Nunavut (see figure 2.1). Nestled near the northeastern end of Baffin Island, this remote, treeless, and glacier-capped island was named for the arctic explorer Robert Bylot. The island lies just south of Lancaster Sound and Devon Island. The better-known and larger islands of Ellesmere and Axel Heiberg are found farther to the north. The island has no permanent settlements but sits across from the community of Pond Inlet, on the other side of Eclipse Sound. Bylot Island is an important wildlife refuge and a key nesting area for snow geese. This island is very similar to Spitzbergen in its stark and rugged landscape but lies slightly farther to the south. However, during the Late Cretaceous, Bylot Island was very nearly at the same paleolatitude as Spitzbergen, both of which were at a lower paleolatitude than today, a paleolatitude nearer to that of the dinosaur-bearing rocks found in the Peel River drainage, Yukon Territory, based on dinosaur-bearing rocks found there.[8] In the Late Cretaceous, Bylot Island had a much warmer and more equitable climate than today. It didn't host the dry deep frigid cold winters it does now, and it supported trees and vegetation similar to what you would find in the cool to warm temperate climate zones of the New England coast to southern Atlantic Seaboard of today. Bylot Island is just outside the southeastern margin of a huge, ancient, structural, sedimentary basin called the Sverdrup Basin. Even though this basin is dominated by marine sediments, the margins contain a shallow marine shelf and massive deltaic deposits that were, like Spitzbergen, a battleground between ocean margin and distal land environments. The basin with its carbon-rich sediments has generated significant natural gas and oil deposits.[9]

A cursory look at a geologic map of this part of the Canadian Arctic would indicate little chance of finding dinosaur-age rocks. The immediate area and Bylot Island are dominated by very ancient Precambrian rocks that are billions of years old. However, a combination of geologic and

Fig. 2.3. First dinosaur footprints (ornithopod) documented in the Arctic. View of almost vertical Festningen sandstone, Helvetiafjellet Formation (Lower Cretaceous), Grøfjorden, Spitzbergen. *Credit: Albert F. de Lapparent, LaSalle-Beauvais Collection.*

climatic history combined to allow for a relatively ice-free faulted wedge of Cretaceous age rocks to be exposed in the southwestern part of Bylot Island.

During the Late Cretaceous, a thick sequence of fine-grained sediments formed just offshore, along the inner shoreline and along the fingers of large, advancing deltas. These sediments often trapped dense vegetation and both marine and terrestrial vertebrate remains. A juvenile hadrosaur metatarsal foot bone picked up in 1987 by Joshua Enookalook, an Innuit native, represented the first evidence of dinosaurs in the Canadian Arctic Islands and kicked off a series of expeditions to this improbable dinosaur venue.[10] Prior to this, the quest for dinosaurs in the Canadian Arctic Islands had focused on the larger islands of Ellesmere and Axel Heiberg.

Joshua was an assistant to Elliot Burden from Memorial University of Newfoundland, who was studying the geology of Bylot Island. His discovery reignited interest in the eastern Canadian Arctic by what was fast becoming the most ambitious dinosaur-hunting program ever put together,

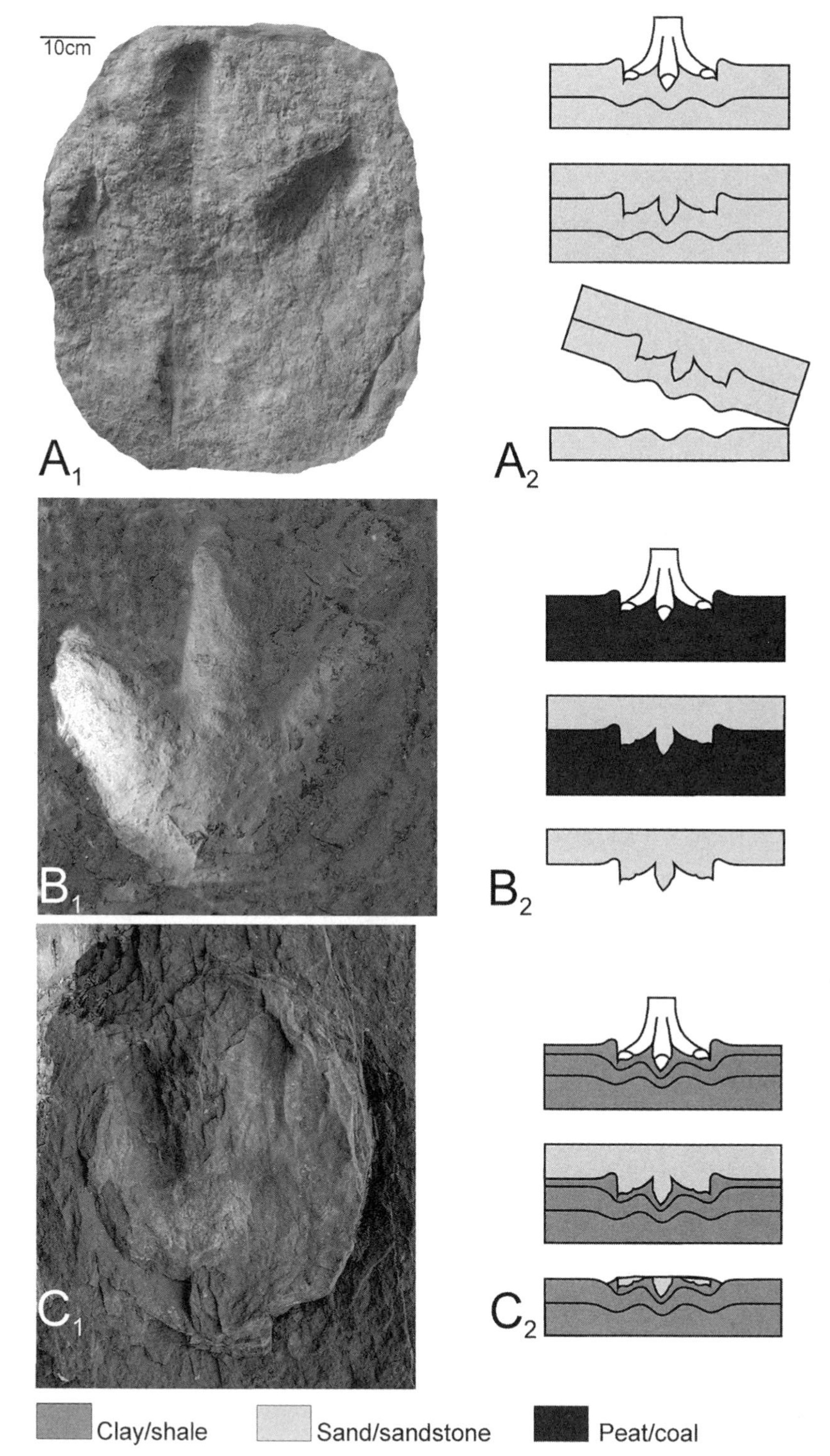

Fig. 2.4. Images of some of the first dinosaur footprints recorded by Albert F. de Lapparent from Svalbard accompanied by a schematic series of possible ways each footprint was made. *Credit: Jørn H. Hurum (Hurum, Milàn, Hammer, et al. 2006).*

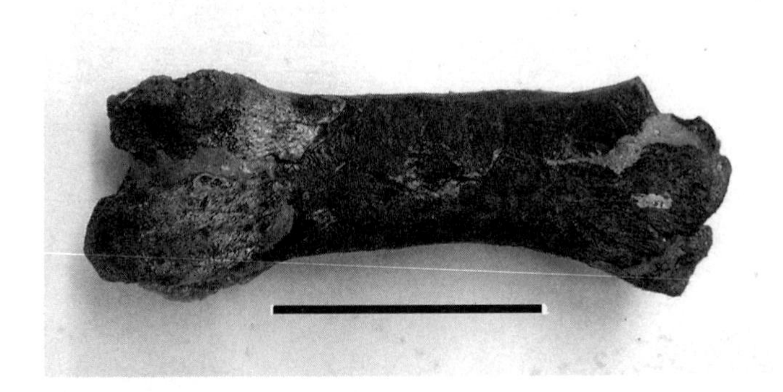

Fig. 2.5. First hadrosaur bone (juvenile metatarsal) found on Bylot Island, Canadian Arctic. Scale = 10 centimeters. *Credit: Canadian Museum of Nature.*

the Canadian-Chinese Dinosaur Project. In 1989, an international team of paleontologists and geologists mounted an expedition to Bylot Island. Although their finds were only fragmentary, these remains represent a fairly diverse dinosaur fauna. The most complete remains are reservedly attributed to two types of hadrosaurs or duckbills (a lambeosaurid, or "crested" duckbill, and a hadrosaurid, or noncrested type). A small theropod, or carnivorous dinosaur, possibly attributable to the family Tyrannosauridae, was also discovered. A total of nine bones have been assigned to dinosaurs and now reside in the Canadian Museum of Nature. In addition to the dinosaur remains, parts of two different types of marine reptiles, a mosasaur and a plesiosaur, were collected. Bones of an extinct type of marine diving bird named *Hesperornis*, similar to a modern loon, were added to this diverse fossil collection. This evidence of an aquatic bird was particularly exciting, for bird fossils are relatively rare, and birds are now considered members of the Dinosauria clade (lineage). Bird fossils give the paleoecologist access to a part of the ancient ecosystem that is seldom available.

Axel Heiberg Island

Axel Heiberg Island in the Canadian Arctic Archipelago is a little over 6° N of Bylot Island. Nestled up close to the west-central waist of Ellesmere Island and separated from it by Eureka Sound is May Point (see plate 1 and figure 2.1). Just to the west of May Point are outcrops of Upper Cretaceous marine rocks assigned to the Kanguk Formation that contain marine microfossils and boney scraps of marine reptiles. Collected along with these marine fossils was a single, abraded, posterior, dorsal, vertebral centrum (main vertebral body). Amazingly, the centrum turned out to be the partial vertebra of a young or subadult hadrosaur.[11] At over 79° N, this is the northernmost place in which dinosaur remains have been found in the present Arctic. The significance of this find goes far beyond its holding the northernmost crown. It supports earlier predictions that the widespread Kanguk Formation should be thoroughly prospected for dinosaur remains and that detailed paleoenvironmental analysis of this rock package that extends throughout the eastern part of the ancient Sverdrup Basin could help to produce a much more complete record of high-latitude dinosaur faunas and associated vertebrates. These finds also promise to reveal a higher resolution picture of Late Cretaceous landmasses that existed in the paleo-Arctic. John Tarduno and his University of Rochester team have added significantly to the nonmarine vertebrate fossil record from Axel Heiberg Island since the earlier expeditions of the 1990s. Collections include fish, a champsosaur (crocodile-like extinct reptile), and three kinds of turtles.[12]

Northwest Territories

Thus far, only a few remains of dinosaurs have been reported from the vast Northwest Territories of northwestern Canada in the District of Mackenzie. The first dinosaur remains were discovered along the East Little Bear River.[13] These dinosaur remains include the articular end of a right quadrate (skull bone) and fragments of a frill attributed to an indeterminate Ceratopsid. Recently, remains ascribed to a hadrosaur have been found by David Evans.[14] All of these remains were found in the Late Cretaceous (Maastrichtian) part of the Summit Creek Formation.[15] This formation is comprised of alluvial conglomerates, sandstones, and interbedded volcanic tuffs and lignitic coals. These alluvial sediments interfinger with shallow marine sediments. The paleoenvironments represented in these rocks point to a great potential for further discoveries. Coal and petroleum finds thus far have opened this once remote area to economic activities that hold promise of better access in the future, especially along the western edge of the Mackenzie Plain and in the Brackett Basin.

Although these fossils are presently found in the upper subarctic at 64.5° N, they represent dinosaurs that lived in the paleo-Arctic somewhere between 68 and 80° N.[16]

Northern Yukon Territory

Only a few fragmentary skeletal remains of dinosaurs have been found in the northern Yukon Territory thus far. These consist of a single distal caudal vertebral centrum and two phalanx fragments from digit V of a juvenile hadrosaur. The centrum is fairly diagnostic of this taxon, but the phalanx pieces are more problematic. Recently, Grant Zazula and David Evans found a humerus that they attribute to a basal ornithopod such as a thescalosaur.[17] All of these fossils were found in the Late Cretaceous (Maastrichtian) lower Bonnet Plume Formation that consists of alluvial sandstones, mudstones, and interbedded lignite. This part of the Bonnet Plume Formation represents an alluvial plain environment similar to the Prince Creek Formation on the North Slope of Alaska. These rocks are now part of the Arctic at 66° N and were at a paleolatitude of between 68 to 80° N during the Late Cretaceous.[18]

These dinosaur skeletal fossils are enhanced by abundant tracks and trackways found to the south near Ross River.[19] These finds are located at the northern and southern ends of two coal- and petroleum-rich structures: the Bonnet Plume Basin to the north and the Whitehorse Trough to the south.[20] The sediments and paleoenvironments contained in these structures should hold an even richer dinosaur record than has been found to date. Again, the economics of carbon-based fuels will bring more government attention and geologic investigations to these areas in the near future.

Fig. 2.6. Axel Heiberg Island, Canadian Arctic Archipelago north of Agate Fjord. *Credit: Leslie Mentel (photo provided by John Tarduno).*

Fig. 2.7. Axel Heiberg Island. University of Rochester campsite with tents on left. Strata of fossiliferous Kanguk Formation in background. *Credit: Leslie Mentel (photo provided by John Tarduno).*

Eastern Siberia and Kempendyay-Teete

The pursuit of Arctic dinosaurs takes us to the huge physiographic region of eastern Siberia and eventually to Koryakia in northeastern Siberia.[21] This is a land that reaches out and falls just short of touching North America—that is, if you ignore the continental crust that lies a few hundred feet under the waters of the Bering Strait. The northeastern corner of this part of Eurasia was intermittently connected to North America by groups of volcanic islands or extensive areas above sea level. This land connection acted as an important evolutionary crucible for the Eurasian and North American flora and fauna for over a hundred million years (see figures 1.2 and 2.1).[22]

Located in one of the more remote areas of eastern Siberia is a dinosaur site known as Kempendyay-Teete (Teteh). This locality is presently just below the Arctic Circle but was above the paleo-Arctic Circle when the dinosaurs roamed this part of Eurasia.[23] Hundreds of miles of travel are required to reach the nearest city or large settlement. The city of Yakutsk is nearly 350 miles (567 kilometers) to the southeast and Mirny, a world-famous diamond mining center, is some 200 miles (325 kilometers) to the northwest. Like most Arctic sites, the locale presents daunting logistical challenges to field parties as a group and also tries the stamina of the members of the team. The site is found on a small tributary to the Vilyuy River and has produced remains of dinosaurs from the Late Jurassic and possibly Early Cretaceous (150–140 million years ago). The Vilyuy is the largest tributary of the mighty Lena River, the Lena being the second longest of the Siberian rivers that flow northward into the Arctic Ocean. Both the Vilyuy and Lena are found within the Sakha (Yakutia) Republic that is part of the Far Eastern Federal District of the Russian Federation. The site is designated as either the Kempendyay-Teete or Teteh (figure 2.1) locality. Kempendyay is a settlement on the Kemp'endyayi River and Teete is a creek that is a tributary to the Botomoiu River that eventually flows into the Vilyuy River.[24] Maps indicate that there are two settlements nearby, Agadary and Suntar. My experience along the Lena and Aldan rivers to the east would make me question whether these are presently inhabited. Many such places on the map in Siberia turn out to be abandoned gulag (prison) camps. This area is most easily reached by helicopter, a mode of transport that is often prohibitively expensive for present-day paleontological expeditions but was much more widely available during the Soviet era. Kempendyay-Teete can now be partially reached by a primitive road from Mirny, but getting all the way there still requires an additional trek of several days on foot or horseback, unless you have access to a surplus tracked vehicle.[25]

The geology and topography of these two sites have much in common with the dinosaur-rich Colville River drainage of Alaska that is described in chapter 3. The Vilyuy-Lena Basin is characterized by broad river valleys, or floodplains, that are underlain by horizontal to gently folded continental sedimentary rocks. The broad floodplains are separated by gently sloping ridges and undulations that are remnants of earlier erosive cycles. Pronounced bluffs are only found along the margins of the Vilyuy River. Like Alaska's Arctic Coastal Plain, an aerial view reveals a myriad of shallow thaw lakes underlain by permafrost with networks of small streams that form a

Fig. 2.8. Overview of Vilyuy River basin with thaw lakes from near Teete site, Yakutia (Sakha) eastern Siberia. *Credit: Pascal Godefroit.*

Fig. 2.9. Teete fossil quarry with taiga forest and melting permafrost, Yakutia (Sakha) eastern Siberia. *Credit: Pascal Godefroit.*

Fig. 2.10. Field vehicle (converted Soviet armored personnel carrier) in Vilyuy River basin on way to Kempendyay-Teete site, Yakutia (Sakha) eastern Siberia. *Credit: Pascal Godefroit.*

"fine silvery web" between many of them. Unlike Alaska's Arctic Coastal Plain, which is dominated by low tundra vegetation, the Vilyuy-Lena Basin is dominated by a larch taiga forest.

Although the fossils are fragmentary, they represent, next to the remains found in Alaska and Koryakia, one of the most diverse nonavian dinosaur assemblages known from the paleo-Arctic. Prior to the discoveries in Koryakia, the Kempendyay-Teete site was the most northerly of dinosaur sites in Siberia.[26] An expedition in 1960 collected the first dinosaur remains. Fieldwork in 1988 and 2002 added significantly to these collections. The fossils of both herbivorous and carnivorous dinosaurs are now in collections at the Paleontological Institute in Moscow. Isolated teeth, toe bones, claws, vertebrae, partial ribs, and an array of miscellaneous partial bones make up the rest of the dataset. The difficulty of working with isolated teeth and incomplete limb elements, such as toe bones, is reflected in the various published attempts at identifying specific taxa such as genera and species. Some of the herbivore teeth have been assigned to an ankylosaur or a stegosaur. Teeth indicative of a large sauropod were recently reassigned to a euhelopodid rather than a camarasaurid sauropod.[27] Teeth indicative of a large theropod appear to be closely similar to those of *Allosaurus*, the famous "tiger of the Jurassic." Smaller carnivores, such as dromaeosaurs, may be represented by sickle-like claws. A Late Jurassic-age date for the remains is fairly certain, but some of the fossils may represent the Early Cretaceous.[28] What is clear is that this locality is highly significant because of its age range and location, and it deserves much more mapping, collecting, and analysis.

Koryakia

The spotlight is now cast on a site that is just a bit farther north and quite a distance to the east of Kempendyay-Teete. This locality is on the Kakanaut River (figure 2.1) in the Koryak Upland area of Koryakia, which is a political division of Khamchatka Oblast. More importantly, it is of very Late Cretaceous age, and during this time it was situated between 70° and 75° N. In addition, this locality is geographically and faunally very closely related to the dinosaur-rich region of Arctic Alaska. Dinosaur bones and other fossils were first collected at this site by Lev Nessov and Lina Golovneva, an extraordinary husband and wife team. These two scientists, he a vertebrate paleontologist and she a paleobotanist, are recognized for their pioneering investigations in some of the most challenging field areas of the former Soviet Union. They and their assistants exhibited a tenacity and dedication to their science that defies adequate description. Persisting with little monetary or logistical support, Lev and Lina often lived out of a suitcase and traveled about on public transport or hitched rides with locals. Based in Leningrad (St. Petersburg), they discovered and described some of the first fossil vertebrates from places like Kasakhstan and Koryakia.

In 1989, Lev and Lina found scattered dinosaur bones and teeth in part of a thick sequence of continental and marine volcano-sedimentary rocks that crop out along the banks of the Kakanaut River in the southeastern part of the Koryak Upland near Lake Pekulneyskoje.[29] These early fragmentary finds were interpreted as being out of place, having been caught up in submarine slide deposits formed on a continental slope. Hadrosaurids (noncrested duckbills) and, very importantly, the teeth of a troodontid and a larger theropod were identified. Tragically, before Lev and his wife could return to conduct more extensive studies, Lev died. Lina still carries on with the passion for science and discovery that she and Lev shared for so many years.

Lina and colleagues from Russia, Belgium, and France have compiled quite a paleofloral record from the Kakanaut Formation in the Koryak Upland. Ten gymnosperm taxa and thirty angiosperm taxa represent the richest megafloral assemblage of Late Maastrichtian age thus far documented for the Arctic. This rich floral record is joined by a diverse assemblage of dinosaur remains collected from a recently discovered microfossil-rich lens in the middle of the Kakanaut Formation.[30] The dinosaur remains consist primarily of assorted teeth, but a few bones have been recovered as well. The bones and teeth have been attributed to a hadrosaur, a neoceratopsian, a nodosaurid ankylosaur, a basal ornithopod, and three theropod taxa. The theropod tooth assemblage closely resembles that found on the North Slope of Alaska (see figure 2.1). The most exciting and unusual dinosaur remains found by this international team is eggshell attributable to hadrosaurids and nonavian theropods—the first record of dinosaur eggshell from the Arctic.

This team of scientists plans to expand this highly significant dinosaur dataset for the Koryak Upland and add to our understanding of this very critical time (Late Maastrichtian) in the history of dinosaurs and the paleo-Arctic.[31] These discoveries in the Koryak Upland in combination with the dataset that has been accumulated for Arctic Alaska over the last twenty-three years present a formidable challenge to several long-standing hypotheses regarding paleo-Arctic dinosaurs and the great end-Cretaceous extinctions that swept up all of the nonavian dinosaurs. This is a subject I further expand on in chapter 3.

Let us now turn our attention to the region that hosts the richest accumulation of paleo-Arctic dinosaurs that has thus far been documented for either of the polar regions—Arctic and subarctic Alaska.

3 A BLACK GOLD RUSH SETS THE STAGE FOR DISCOVERY IN ALASKA

If you have seen the icon for the Sinclair Oil Company, a silhouette of a four-legged long-neck dinosaur, then you may have concluded that petroleum and dinosaurs go hand in hand. The misconception that petroleum is derived from dinosaurs is still quite prevalent even among those who are not familiar with the Sinclair symbol. Physicists and chemists in the 1800s considered petroleum to be nonbiogenic and concluded that petroleum was a residue of the formation of the Earth. This hypothesis was discarded by the 1950s and replaced with a theory that petroleum had a biogenic origin, but there was no consensus on the specific types of organisms that were transformed into oil. Geochemical research over the last three decades has focused on plankton and microorganisms as the biomass that combines with rock-forming processes to end up as the flame on your stove or the gas in your tank. It is now clear that most petroleum is derived from ancient planktonic microorganisms rather than dinosaurs and their vertebrate kin.[1] Still, interestingly, there is an important and valid connection between the discovery of oil and dinosaurs—especially in Alaska.

Alaska's Dinosaurs and the Search for Oil

When the tale of the discovery of Alaska's dinosaurs is recounted, the role of oil exploration takes center stage. For it was a field geologist engaged in the search for oil during the 1960s who collected and recorded fossils that would later be identified as dinosaurian. Petroleum exploration and development that goes back to the early nineteenth century in Alaska is what ultimately set the stage for this exciting and unexpected discovery of dinosaurs. Oil seeps and other surface finds were reported from several parts of Alaska in the 1800s, including the Alaska Peninsula and the Arctic Coastal Plain. From the early 1900s to the 1950s, subsurface discoveries spurred on further efforts along the northern Alaska Peninsula and the Yakataga area in the southeastern edge of the Gulf of Alaska. Several large oil companies were involved, but most of the oil fields were found by independents, also known as wildcatters. But with the nationalization of the Iranian oil industry, which in the early 1950s was still controlled by the British Anglo-Iranian Oil Company, the days of independent oil drillers in the Arctic would come to end. Following the assassination of Iranian prime minister Ali Razmara in 1951 and much political unrest in Iran, Iranian petroleum resources were nationalized, and Anglo-Iranian Oil morphed into British Petroleum (BP). The loss of its monopoly in Iran forced BP to look elsewhere for oil. It extensively analyzed potential petroleum-rich areas throughout the world; recognizing important similarities between the oil-rich Zagreb foothills of Iran and Iraq and those of the Brooks Range, it eventually made its way to Alaska's Arctic and was soon followed by Shell, Chevron, and several other oil companies. Subsequently, in the 1960s, the much smaller Richfield Oil Company would join BP and the others in northern Alaska. In a bizarre twist of fate, Richfield would become the Atlantic Richfield Company (ARCO) and would then make the biggest discovery in what would be called Alaska's "North Slope oil bonanza."[2] Oil exploration in Arctic Alaska during the 1960s and '70s brought forth a new label for this region and a new term to the worldwide petroleum lexicon—North Slope. This appears to have grown out of general references to the area as the north slope of the Brooks Range and then more expansively as "Alaska's North Slope." The region is now almost universally referred to as just the "North Slope."

The Brooks Range is a spectacular topographic boundary, delimiting the Arctic Coastal Plain (see figures 1.2 and 3.1) that spreads north from the foothills of the Brooks Range to the scalloped edge of the Beaufort Sea and the Arctic Ocean. Although the Brooks Range seemed the most likely place to explore for oil because it held the kinds of rocks and earthly contortions that can act as traps for crude oil, it did not prove to be the treasure trove that it first appeared to be.[3] Like many aspects of Alaska, the range was not what it seemed at first glance. The oily treasure instead turned out to be nested far below the nearly flat, ice-locked sediments piled up by millennia of river meandering and flooding that is technically labeled by geologists as the Arctic Coastal Plain.

The Arctic Environment of Alaska

The Arctic of Alaska is comprised of three major physiographic provinces.[4] These are the Arctic Mountains, which include the central and eastern Brooks Range, the Arctic Foothills, and the Arctic Coastal Plain. It is the Arctic Coastal Plain that, thus far, has produced the greatest concentrations of petroleum in Alaska and North America. More importantly for the main subject of this book, it also contains the greatest known concentration of dinosaur remains in the Arctic and Antarctic combined (figure 2.1).[5] This huge province encompasses over 89,000 square miles, an area slightly larger than the state of Minnesota. It is one of the least populated areas of Alaska with only a little over nine thousand residents.[6] The boundaries of the Arctic Coastal Plain of Alaska are partially defined by the presence of continuous permafrost (see the glossary). Geographically, the Arctic Coastal Plain extends from the northern edge of the foothill belt of the Brooks Range to the shores of the Chukchi and Beaufort seas, subdivisions of the Arctic Ocean. This area is a harsh and unforgiving land often described as desolate, forbidding, and colored by a "dismal gray" or a depressing white during its long winter. Journalist Joe McGinniss refers to it as "that part of North America which could make you believe that the earth was indeed flat and that, at last, you had come to its edge."[7] It is

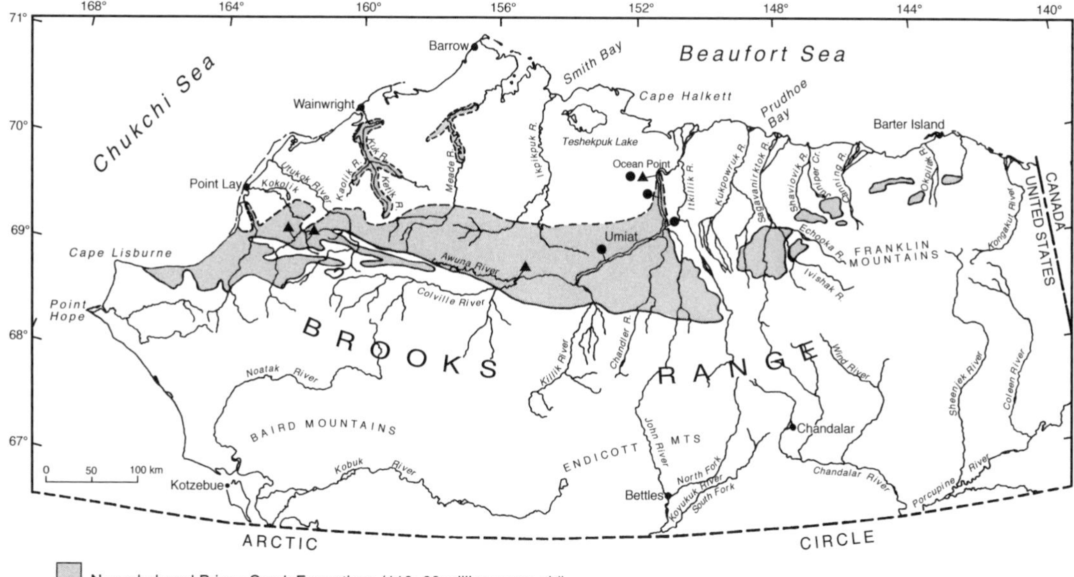

Fig. 3.1. Map of North Slope showing distribution of the Nanushuk and Prince Creek formations, dinosaur localities, and important geographic features. *Credit: Gil Mull and Curt Huffman Jr., modified by Dixon Jones from U.S. Geological Survey maps, following Huffman (1985).*

an environment marked by seasonal extremes. Winter brings darkness or twilight, howling winds, and deep subzero cold that caps the landscape with a thin mantle of snow, except where muted topography traps it as it drifts across the seemingly endless plain. Lakes and rivers freeze to depths of several feet, and the survivors of the last "ice age" continue to scratch out a meager living. The sun sinks just below the horizon for nearly two months of winter twilight. Summer is highlighted by a sun that doesn't set except for a couple of hours, and for nearly two months this land is host to an outbreak of plant and animal life that is best described as frenetic. What was a sea of monotonous white changes to an ocean of green that melts into the blue of the horizon, broken by countless thaw lakes and occasional pingos (frost-heaved mounds).

Courses of meltwater flow down the northern flank of the glacier-scarred Brooks Range and converge to form a few mighty rivers that seek the northern sea. The larger rivers scour out steep bluffs along some parts of their march to the sea. These offer the only direct windows into the geologic past for the field geologist. As the days lengthen, the tundra springs forth in verdant splendor and gives rise to multitudes of biting and sucking winged insects. Researchers have found that as many as nine thousand mosquito bites per minute can be recorded during the summer in some parts of Alaska. Caribou are particularly vulnerable and can be seen to run about wildly when the biting insects reach their peak during an early part of their hatching cycle. Some reports claim that caribou can be driven to run until they drop dead if they cannot reach a deep enough stream or river in time.[8]

The dense vegetative mat masks the geologic trends that stretch between rivers. The swarms of insects harbored by this tangled mat can test the sanity of humans who attempt to traverse it. The tundra flats present

Fig. 3.2. Aerial view of thaw lakes and polygonal ground on North Slope with Colville River on the horizon. Image taken from rear ramp of a Chinook helicopter. *Credit: Phelana Pang and Christopher Tolentino.*

Fig. 3.3. Volunteer Bill Hopkins surrounded by tundra at the top of Colville bluffs. Note pingos near horizon on left. *Credit: Roland Gangloff.*

the hiker with a nearly impassable combination of tussock mounds and ankle-threatening rivulets. The threads of the geologic puzzle that can be gathered with hammer and eye are sewn together with the use of aerial photos, seismic profiles, and well cores.

Temperatures in the summer often climb the Fahrenheit scale to the 70s and 80s and can even reach the 90s. Great brown bears cruise the tundra and join the herds of caribou and musk oxen trying to fatten up before the long winter returns. Arctic ground squirrels seem to pop up everywhere. Many of these tunneling rodents disconcertingly undermine the tops of the steep river bluffs with their subterranean "cities." These underground labyrinths, in turn, are often torn asunder by cruising brown bears that pop the sizable inhabitants into their mouths as if they were candy corn. These faunal machinations, combined with melting ice wedges and lenses, make climbing on or working below the bluffs especially hazardous. These are just a few of the many challenges that summers on the Arctic Coastal Plain hold for field geologists and other scientists.[9]

Alaska's Dinosaurs Discovered: The Role of Oil Exploration and Field Geologists

The realization that the rocks and structures of the Brooks Range held only tantalizing caches and not the commercial amounts of the black treasure the oil companies sought spurred a geological march down the northern slope of the Brooks and on toward the Beaufort Sea (figure 3.1) that brought forth the first documentation of dinosaurs in Alaska as well as the largest oil strike and subsequent economic boom in the history of the United States.[10] Robert Liscomb has been given credit for the discovery of the first dinosaur remains in Alaska, which was part of a thirty-seven-year journey.[11] He came to Alaska as an oil exploration geologist for Shell Oil Company after earning degrees from Stanford University in California and Frederick Wilhelm University in Bonn, Germany. His ten years with Shell had taken him to the Rocky Mountains, the Salt Lake City area, and finally, in 1961, to Alaska. The 1960s witnessed the beginnings of the "black gold" rush that eventually led to the extraordinary petroleum discoveries near Prudhoe Bay in northern Alaska. Robert Liscomb was assigned to map rock exposures along the Colville River as part of Shell Oil's effort to evaluate the potential for oil riches above and below ground on Alaska's Arctic Coastal Plain.

Liscomb was dropped off, sometimes alone, by helicopter or fixed-wing aircraft for several days before being picked up at a prearranged landing site. The personal challenges required of those who conduct fieldwork in Alaska's vast and remote North Slope are hauntingly portrayed in poetry found among Liscomb's papers after his death:[12]

Fig. 3.4. Photo of Robert Liscomb in his thirties. *Credit: Photo provided by Bonnie Liscomb Brundage.*

> Snow-peaks and deep gashed draws corral me in a ring
> I feel as if I was the only living thing on all this blighted earth;
> And so I frounst and shrink and crouching by my hearth
> I hear the thoughts I think day after day the same,
> Only a little worse no one to grouch or blame—
> Oh, for a loving curse! Oh, in the night I fear,
> Haunted by nameless things, just for a voice to cheer,
> Just for a hand that clings!
> I will not wash my face, I will not brush my hair,
> I "pig" around the place—there's nobody to care,
> Nothing but rock and tree, nothing but wood and stone,
> Oh, God it's hell to be
> Alone, alone alone.

Fig. 3.5. Some of the first hadrosaur bones from Liscomb Bone Bed collected in 1961 by Robert L. Liscomb. A, B, E, F are pedal phalanges (toe bones); C and D are caudal vertebral centra. *Credit: Kyle Davies and Texas Memorial Museum, Austin, Texas.*

Field geology can be a very lonesome and dangerous pursuit, since it is not always possible to have a field assistant. Mapping and collecting rock formations in the Arctic is especially demanding and a person who undertakes such work must be physically and mentally very strong. It is interesting to note that Robert Liscomb was preceded, on the North Slope, by several pioneering mapping geologists. Earnest Leffingwell, a truly legendary Arctic field geologist, was responsible for making the earliest detailed geologic maps of much of the North Slope between 1907 and 1921. He did it mostly alone—a herculean accomplishment that resulted in the definition of the most important oil-producing rock unit. These maps laid the foundation for the world-class petroleum discovery at Prudhoe Bay.

As a well-trained mapping geologist, Liscomb took both rock and fossil samples. He marked the samples with codes, recorded them in his field book and then sent to Shell Oil Company labs for further analysis. These samples would allow other geoscientists, such as geochemists, at Shell to further evaluate the hydrocarbon-bearing potential of the rocks and pin down more precise geologic ages for the rock packages that he was measuring and tracing across the country. Those samples deemed important and worthy of the expense would ultimately end up in a central company repository. Such samples were also deemed proprietary and kept secret from the "outside" world and, especially, from other oil companies. Unbeknownst to Liscomb, he had included dinosaur bones in with collections of younger mammal bones during his mapping along the Colville. This collection site was not far from Ocean Point (figure 3.1), one of the few locations that is named on topographic maps of the area.

Most oil or mineral exploration geologists are not well acquainted with vertebrate fossils, and they rely upon vertebrate paleontologists to evaluate them when they find them. In most cases, vertebrate fossils are noted but not collected unless they might make an attractive mantelpiece or doorstop. Invertebrate fossils, such as clams, snails, ammonites, and trilobites, are usually more familiar to economic geologists, who find them useful for determining the refined ages and environmental settings of the rocks that they map.

Liscomb was unusually thorough in his collection of field data, and this attention to detail led to one of the most important dinosaur finds in the Arctic and North America. Tragically, Liscomb never knew of his important contribution to dinosaur research in the Arctic. He was found floating face up in a small cove at the base of a 100-foot cliff on a small, remote island in the Prince Williams Sound, some 80 miles south of Cordova and over 900 miles to the south of the Colville River. He apparently died in a rock fall while conducting field investigations and succumbed to severe multiple fractures of his skull, limbs, and trunk.

None of the field geologists from the state or federal government or from other oil companies, before and over two decades after Liscomb's work, reported the presence of dinosaur bones and teeth from outcrops along the Colville River. These fossils accumulate by the hundreds each year along the base of the river bluffs that include what is now called the

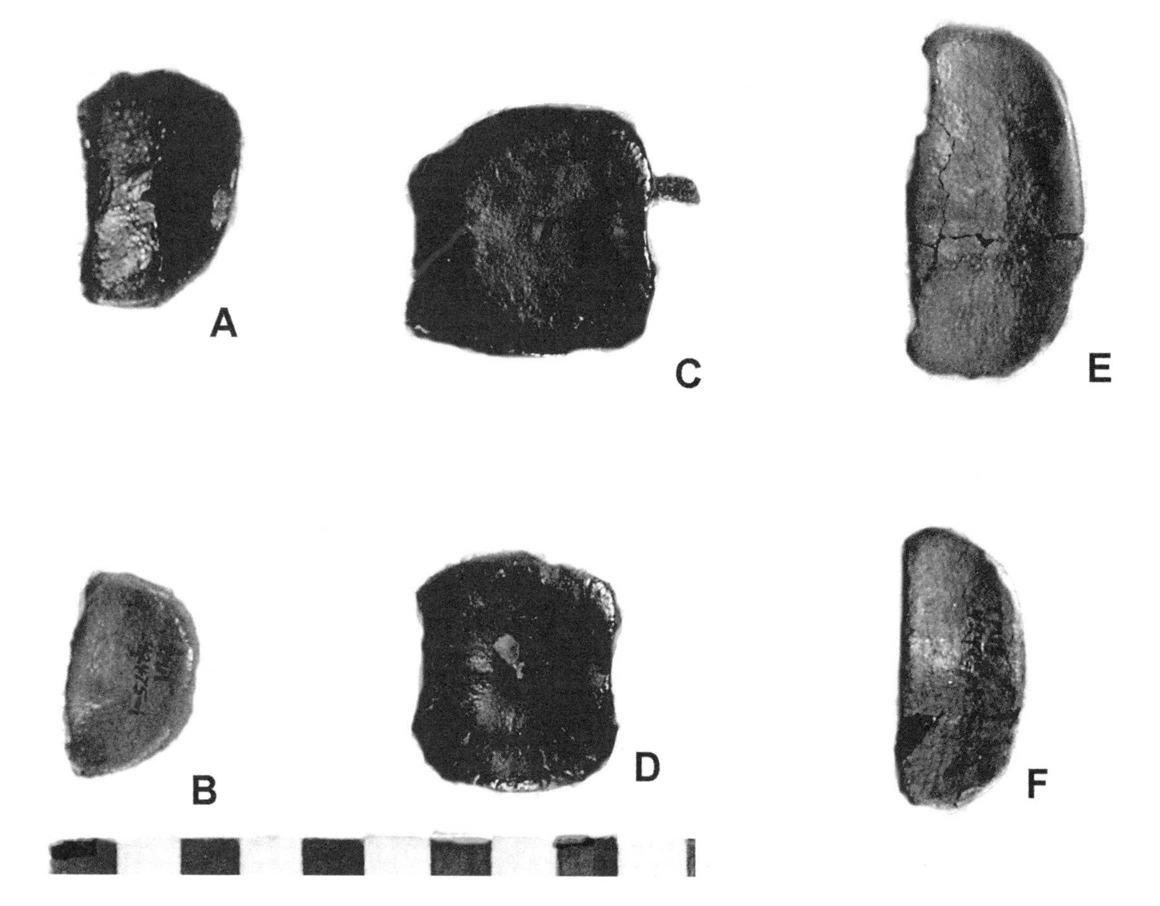

Liscomb Bone Bed. So Liscomb deserves to be posthumously congratulated on his dedication and conscientiousness as a field geologist and scientist.

The First Dinosaur Bones are Rediscovered

In 1983, paleontologist Richard Emmons was assigned to a review of the Colville area by Shell Oil Company. Coincidentally, Emmons had accompanied Robert Liscomb in the field on the day that Liscomb died . Emmons compiled all of the geologic data, including all of the archived fossils. He came across the bones that Liscomb had collected and identified some of them as representing fossil marine mammals. Since the specimens were no longer considered to be proprietary, he shipped them to Charles Repenning at the U.S. Geological Survey in Menlo Park, California, a well-known paleontologist who specialized in marine mammals. Repenning verified that the shipment did indeed contain some fossil marine mammal bones, but he also recognized dinosaur bones that he tentatively attributed to hadrosaurs.

Struck by the Alaskan venue of the bones, Repenning sent the hadrosaur bones to Wann Langston Jr. at the University of Texas, Austin, for further evaluation. Langston quickly determined that the collection represented several individuals and assigned his assistant, Kyle Davies, to work out further details. Davies published the first photos and descriptions of the bones and offered a preliminary interpretation of their significance.[13] Understanding the importance and implications of the find, Langston excitedly reported his results to Repenning and strongly suggested that the U.S. Geological Survey relocate the fossil site (which it had the power to do since the site was located in the National Petroleum Reserve) and further evaluate it. Thus, one of the unintentional results of field exploration for petroleum deposits on the North Slope of Alaska was the discovery of the first dinosaur remains in Alaska, and that would eventually lead to a remarkable record of Cretaceous-age dinosaurs that would surpass that compiled in all of the rest of the polar regions.

The Extinction of the Dinosaurs: The Debate Centers on Alaska's North Slope

During the period from 1983 to 1985, Wann Langston and Chuck Repenning's recommendations that Liscomb's fossil site be verified and that the fossil remains be studied further were followed up by the U.S. Geological Survey. In the summer of 1984, a team comprised of Elizabeth Brouwers, Thomas Ager, and David Carter verified the Liscomb locality and collected more fossils. The results of this fieldwork were used as the basis for further exploration in 1986 by a field party that included a sedimentologist and a specialist in mollusks in addition to Brouwers and Carter.[14] In between these 1984 and 1986 expeditions, a joint expedition principally funded by the National Science Foundation was undertaken, led by paleontologists William Clemens of the University of California, Berkeley, and Carol Wagner-Allison of the University of Alaska, Fairbanks.

How Clemens's work came to be funded by the National Science Foundation is a long story that goes back to 1980, when Walter Alvarez, a Nobel Laureate nuclear chemist, and colleagues, published a seminal paper in the journal *Science* that put forth a new hypothesis that Clemens disagreed with explaining the extinction of the dinosaurs. Prior to 1980, attempts to explain the demise of dinosaurs had fueled a rather esoteric debate that had been primarily based on unrestrained speculation and erroneous assumptions regarding dinosaur biology.[15] Clemens vigorously argued from his experience with the fossil record of mammals and dinosaurs in end-Cretaceous rocks from Montana. He maintained a uniformitarian view and argued for a geologically slow and gradual decline of dinosaur populations ending in extinction at or near the end of the Cretaceous period. Clemens argued that the extinctions that marked the end of the Mesozoic era reflected the impact of an array of long-term causes. These included climate change and marine regression, both of which have been documented in the geologic record.[16] What became known as the Alvarez camp argued for a shorter-term catastrophic view that called for a massive impact by an extraterrestrial object such as a meteorite or bolide.[17] The possibility of such an impact was based on the discovery in marine sediments of unusually high concentrations of the element iridium, too high to have had a terrestrial origin. The lineup of protagonists and the resurrection of catastrophism greatly appealed to the general public and spurred a host of television programs and magazine articles. This, combined with new finds of dinosaurs in both hemispheres, helped to reawaken a latent

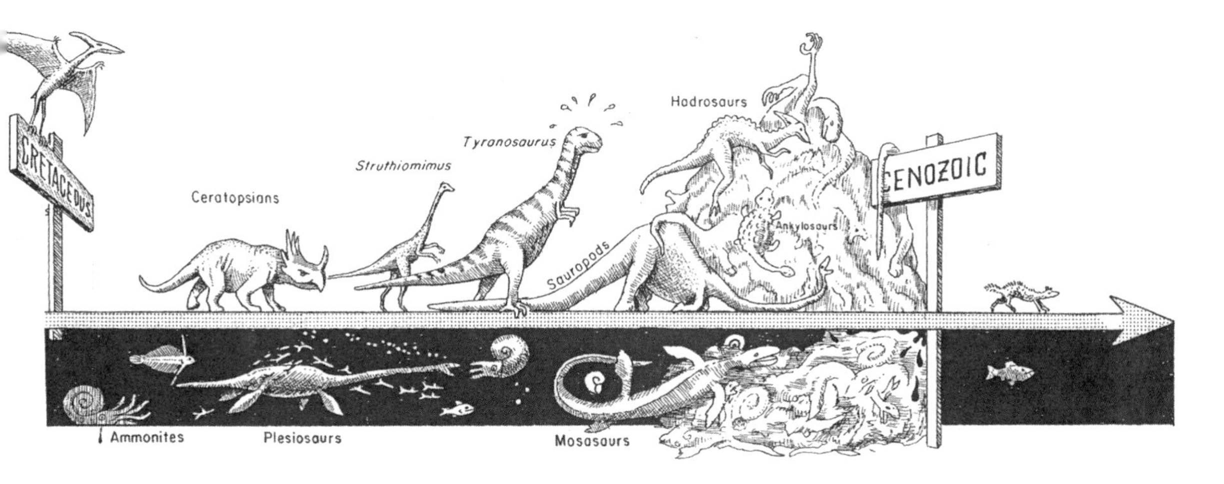

Fig. 3.6. Cartoon of the K-T (K-P) extinctions as envisioned by strict catastrophists. *Credit: John C. Holden.*

interest in dinosaurs on the part of the public. An important shift was taking place. Arguments primarily related to highly speculative work on dinosaur physiology, paleoecology, and behavior were being replaced by those that were based on physical evidence that could be discerned in the geologic rock record.[18] Evidence such as isotopes, the detailed record of marine and terrestrial rocks, the record of volcanic eruptions, and better records of dinosaur fossils were now being focused on. The purpose of this book is not to go into an exhaustive discussion of the dinosaur extinction debate but rather to discuss the role of dinosaur finds in the Arctic as they relate to this interesting debate.[19]

One of the predictions, formulated on the basis of the impact hypothesis, was that the dinosaurs would be subjected to a short-term global winterization of climate for at least a year that would do these "cold-blooded" beasts in. This hypothesis stemmed from the Alvarez camp's work in the nuclear weapons arena and was popularized as the "nuclear winter" scenario.[20] The debate that was engendered by this hypothesis led to a worldwide search for evidence of end-Cretaceous impact craters, impact debris, and iridium-enriched layers as well as new approaches to determining the probable physiology of dinosaurs.[21] With the discovery of the first evidence of dinosaurs in the eastern Arctic island of Spitzbergen in 1961, the possibility of finding abundant and widespread dinosaur remains in the Arctic began to be important to the highly engaging and ongoing debate regarding the physiology and behavior of dinosaurs. Abundant and widespread endemic dinosaur faunas in the paleo-Arctic would strongly challenge the simplistic "nuclear winter" extinction hypothesis since these faunas would indicate that dinosaurs were capable of adapting to the winter cold at high latitudes (the winters, as noted in chapter 2, would not have been as frigid as today; the climate would have been comparable to maritime temperate climates of the western and eastern coast of North America now).[22] The long-term presence and adaptation to the high paleolatitudes would also counter the hypothesis that dinosaurs were just scaled-up "lizards" and therefore ectothermic ("cold-blooded") and must have been confined to warmer lower latitudes. The "cold-blooded" characterization of dinosaurs and the speculation as to their exact relationship to lizards and other living reptiles had held sway among the majority of vertebrate paleontologists and the general public ever since Richard Owen had erected the name Dinosauria in 1841.[23] The one rediscovery and subsequent new discoveries of dinosaurs in the paleo-Arctic that began on the North Slope of Alaska in the early to mid-1980s eventually led not only to an abundant and widespread record but to a diversity that would surprise and impress many dinosaur specialists working in the lower latitudes.

The period 1983 through 1986 witnessed a substantial growth in the collection of paleo-Arctic dinosaur remains from Alaska and an elucidation of their context. In August 1985, Clemens and Allison described some of the exciting results of their expedition to the Liscomb Bone Bed on Alaska's North Slope at a public forum. This public talk and subsequent interviews at the University of Alaska Museum in Fairbanks, where Allison was a curator, triggered a three-month explosion of newspaper and magazine articles. Local and state of Alaska newspapers such as the *Daily News-Miner* and the *Anchorage Daily News* proclaimed that these Arctic finds cast doubt on or at least strongly challenged the most popular dinosaur extinction hypothesis ("death" by meteorite) and buried the image of the dinosaurs as tropical to subtropical creatures that lived in lakes and steamy "jungles."[24] The *San Francisco Chronicle's* noted science writer David Perlman, the *New Scientist*, and *Time* echoed these conclusions concerning the significance of these Arctic dinosaurs from Alaska. It is important to note that the debate concerning the demise of the dinosaurs at the end of the Cretaceous period was still very much alive among scientists and the general public in 1985. In May 1985, *Time* magazine featured a cover and extensive article titled "Did Comets Kill the Dinosaurs?"[25]

The extensive collections of dinosaur remains from the Liscomb Bone Bed combined with the Spitzbergen tracks (see figure 2.3) formed the basis for a solid challenge to the validity of the simplistic "nuclear winter" part of Alvarez and company's impact hypothesis. It was becoming clear that dinosaur remains in the Arctic were too abundant and widespread in time and geography to represent a rare "fluke" or set of stragglers. That there had been endemic Arctic dinosaur populations was now clearly a real possibility. These preliminary results laid a sound foundation for Clemens to approach the National Science Foundation for expanded funding. The year 1987 would see the initialization of a four-year program supported by the foundation, the Museum of Paleontology at the University of California Berkeley, and the University of Alaska Museum in Fairbanks.

1987: North Slope Discoveries Prove to Be Significant

The year 1987 turned out to be very critical to the scientific appraisal of the significance of the finds in Alaska and their challenge to Alvarez's short-term impact-winterization hypothesis. Three seminal papers were published in the scientific literature that, for the first time, spoke to the potential abundance and variety of dinosaurs that might be locked up in paleo-Arctic rocks. The initial paper was the first to identify and illustrate the bones (figure 3.5) that had been collected by Robert Liscomb.[26] Kyle Davies, Wann Langston's assistant at Texas Memorial Museum, identified a partial tibia, two pedal phalanges, and several caudal vertebrae as belonging to a hadrosaur or duckbill dinosaur. The collection also contained several nondefinitive rib fragments.

The second paper summarized the stratigraphy and paleoecology represented by the rocks that yielded the Liscomb bones. In addition, the paper reported the remains of tyrannosaurid and troodontid theropods and also incorrectly attributed the hadrosaurid remains to a "crested," or lambeosaurine, type.[27] The paper asserted that the preponderance of evidence pointed to winter residency and the adaptation to winter darkness on the part of the hadrosaurs. Thus this paper pointed out that several short-term environmental effects ascribed to an impact or volcanically induced winterization "may not have been the direct cause of the demise of the dinosaurs."

The third paper summarized the Cretaceous-age vertebrates collected prior to 1986 and held in the collections of the Smithsonian.[28] These included the first fossil turtles in Alaska and the first ceratopsian dinosaurs based on a partial horn core and an incomplete rear ball joint that articulates with the top of the spinal column. The paper also summarized the stratigraphy and paleoclimatology represented by the rocks and plant fossils found along with the vertebrate fossils and considered the question of whether the accumulated dinosaur fauna of Alaska pointed to overwintering or instead to long-distance migration winter survival strategies. The paper concluded that too little was known about dinosaur physiology, especially energetics, forage requirements, and Arctic winter temperatures, to allow for a clear-cut choice as to the most likely scenario. However the paper did set off a spirited debate in *Science* magazine.[29] This set of responses to this third seminal paper centered on probable temperatures in the Late Cretaceous Arctic and near Arctic based on fossil plant parameters and therefore on what temperatures would dinosaurs have had to adapt to and when.

A fourth paper by Gregory Paul published in 1988 presented an extensive analysis of the climatological, migrational, and physiological implications of Cretaceous polar dinosaurs that showed that the ectothermic (cold-blooded) dinosaur physiologic model was questionable even if these dinosaurs were migrants.[30] Paul also pointed out by way of disputing the migration idea that Arctic dinosaur migrants would have to have covered at least 3,600 miles (6,000 kilometers) each year. Philip Currie followed Paul's discussion with a paper in *Natural History* magazine that argued for long-distance annual migration by Arctic dinosaurs that would cover the distances determined by Paul.[31] Currie called on Arctic caribou as a model or analog. It should be clear that the dinosaur record of the Arctic, and especially Arctic Alaska, was becoming increasingly important to our understanding of dinosaur biology. The summer of 1987 also marked the beginning of my involvement in the excavation, curation, and study of the most abundant and diverse dinosaur fauna ever found in both polar regions (see figure 2.1).[32]

"Fresh" Bones and Creationists: Early Reports Spawn Another Controversy

Our expanding knowledge of dinosaurs and their world, subjects that were becoming very appealing to children and the adult public, spawned another controversy. This controversy was initiated by a segment of the Christian community (biblical fundamentalists). Some fundamentalists propose that there is a conflict between science and Christianity any time that data, hypotheses, or theories cannot be reconciled with the literal interpretation of the Old Testament. These Christian fundamentalists include several subgroups, such as young Earth adherents and creation scientists, and creation scientists include flood geologists and intelligent design proponents.[33]

Some creation scientists continue to misrepresent the early data on North Slope dinosaurs and ignore subsequent facts published in the scientific and popular literature. Since they do this in print and on the World Wide Web, I would like to set the record straight, in print, given that few of my colleagues have done so.

"Fresh Dinosaur Bones Found" is the title of a 1992 article published in *Creation*, a journal of the Answers in Genesis organization. The author, Margaret Helder, holds a PhD in botany and has been described as "probably the most prominent women in creation science."[34] Helder made the erroneous assertion that the dinosaur bones were "fresh" despite the publication of several descriptions and illustrations of the permineralized nature of the bones and teeth found in Alaska during the late 1980s. She apparently misinterpreted statements made by Kyle Davies that described the bones as being "remarkably well preserved." The Davies report also stated that the bones exhibited "little permineralization."[35] Davies's paper is cited in Helder's references. Another 1987 paper by Elizabeth Brouwers and colleagues described the bones from the same locality as "remarkably well preserved.[36]

Margaret Helder may have been a bit misled by these statements regarding the state of preservation of the first bones that were collected from what is now called the Liscomb Bone Bed. However, to equate fresh and well preserved or fresh and lightly permineralized is nevertheless odd. In addition, neither Helder nor these scientists ever spoke to the degree of mineral replacement or modification in the bones. Helder has never retracted or modified her original claim that "fresh" bones were found the Liscomb Bone Bed. Creation websites still repeat this misleading and inaccurate report as originally published in *Creation* magazine, despite several papers published between 1987 and 2001 that described the bones, their color, and degree of permineralization and mineral replacement. Subsequent work by me, some of my graduate students, and my colleague Mark Goodwin, of the University of California Museum of Paleontology, using thin sections and microscopic examination have demonstrated that the degree of permineralization ranges from slight to heavy even along the length of a single bone. Some bones are highly permineralized and have crystals growing into cavities (plate 3). The degree of permineralization and mineral replacement does range quite a bit within individual bones and throughout the enclosing mudstone. All of the bones (approximately six thousand) that I have examined, either megascopically

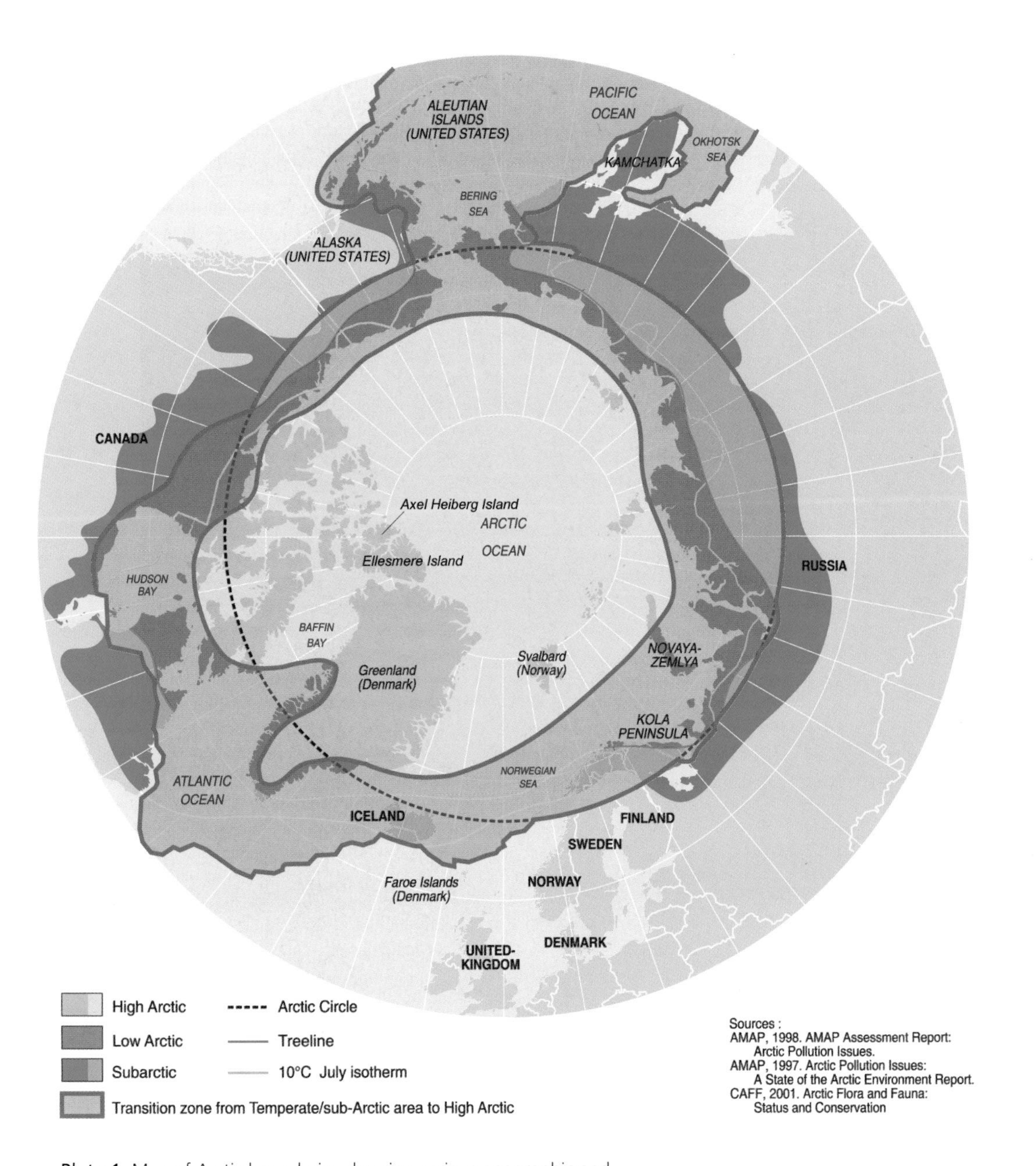

Plate 1. Map of Arctic boundaries showing various geographic and ecologic definitions of the Arctic. *Credit: AMAP and CAFF.*

Plate 3. Fragment of Edmontosaurus limb bone from Liscomb Bone Bed showing mineralization and internal architecture. Zeolite crystals, and nodular masses of francolite, and siderite are visible at left end. Dark brown bone matrix contains francolite, calcite, limonite siderite, and pyrite. *Credit: Roland Gangloff.*

Plate 4. Don Lofgren flanked by recent slides, as he stands on a bone bed just downstream from the Liscomb Bone Bed, 1992. Note white volcanic ash bed that lies above the bone bed. *Credit: Roland Gangloff.*

Plate 2. Facing. *Edmontosaurus* mother and child under the aurora borealis on North Slope, Alaska. *Credit: Karen Carr and Gerald and Buff Corsi/Focus on Nature, Inc.*

Plate 5. Aerial view of Umiat surrounded by thaw lakes and expanses of tundra in blazing fall colors with airstrip and support buildings near center and Colville River to the north and east. *Credit: Rick Reanier.*

Plate 6. Artist's reconstruction of mid-Cretaceous (Albian to Cenomanian) dinosaur assemblage and environment represented by trackways and other fossils near Ross River, Yukon Territory, Canada. In the foreground, two ankylosaurids face a large tyrannosaurid while smaller theropods and ornithopods hunt and forage, in and along, the river's edge. *Credit: George Teichmann and Yukon Heritage Resources.*

Plate 7. Facing. Artist's reconstruction of Late Cretaceous (circa seventy-five million years ago) dinosaur assemblage and environment from the North Slope of Alaska. Foreground from left to right: *Pachyrhinosaurus* and small thescelosaurs, head of large tyrannosaur (possibly *Albertosaurus*), *Dromaeosaurus,* and a saurornitholestid. Background from left to right: *Edmontosaurus, Troodon* chasing a young *Edmontosaurus* with a pachycephalosaur far to the rear. *Credit: Karen Carr.*

Plate 8. Map of present and potential oil, gas, and coal development in the circumarctic. *Credit: AMAP and CAFF.*

Plate 9. Fly in amber from Baltic coast (twenty-three to thirty-three million years old). *Credit: anonymous (enlarged and cropped by the author).*

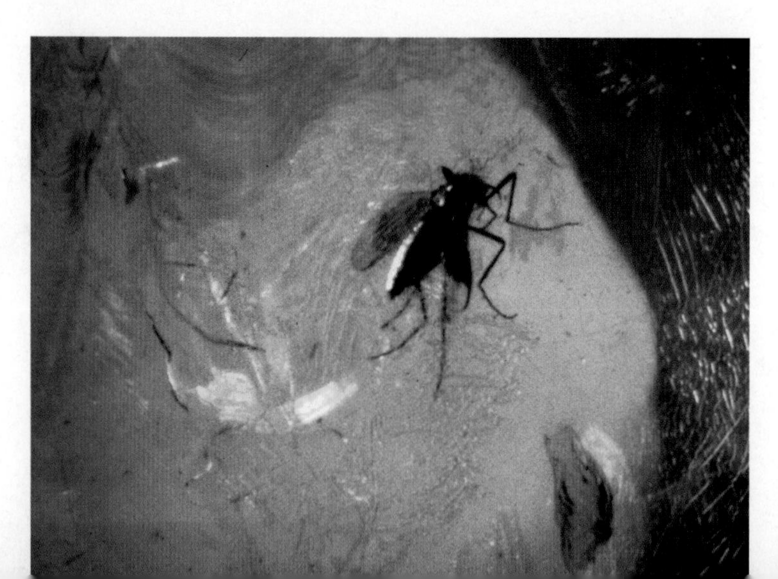

or microscopically, exhibit replacement, usually by silica, and most have some degree of permineralization by several other minerals. X-ray diffraction and microprobe analysis of the bone indicates mineral replacement and modification. Finally, what some of the earliest researchers reported as staining is not simple staining but rather replacement and induration of the original bone structure. Mark Goodwin and his colleagues at the University of California used proton-induced X-ray emission to show that the amount of iron and manganese in the bone architecture were several orders of magnitude greater than in modern bone.[37] The Liscomb bones studied also showed a depletion in the amount of calcium and phosphorous when compared to modern bone. Some of these results are recorded in a publication by Tony Fiorillo and me.[38]

Helder could have elicited a response from the authors as to her conclusion and/or could have gained access to museum collections to view the remains under question prior to publishing her article. But instead she did neither of these things. She merely asserted that the remains from Alaska's North Slope were "fresh" (see plate 3) and that "such bones could never have lasted 70 million years." She went on to further state that the bones are probably only thousands of years old. Finally, she admonished that "it is time geologists recognized the implications of their own data."[39]

Helder's article continues to mislead, as it is posted on several creationist websites. This first article was followed in 2003 by another that distorted and misstated data from research published in scientific papers and from articles in popular magazines on other dinosaur sites from Alaska's North Slope. Entitled "The Big Splash," it is posted on a creationist website in Alberta, Canada.[40] Helder repeated her earlier claims the bones found were unfossilized and committed several gross inaccuracies about the history of discovery and the geographic extent of the bone bed in which the fossils were found. Helder then focused on the possibility that the horned dinosaurs reported from Alaska were drowned in a great flood that was beyond anything we could normally experience. The biblical implications are left up to the reader.

Flood Geologists Take on Alaska's Arctic

The "land of the midnight sun" certainly has attracted a plethora of strange and unusual individuals over the last hundred years, and 1994 may provide one of the stranger chapters in Alaska's history book. In this year, a self-proclaimed scientific expedition and its attached cast of diverse characters undertook a trip to the Artic, declaring that it would drop "a bombshell on the scientific community." Instead, the team not only accomplished little of scientific value but also barely escaped ending up as another cheechako tragedy.[41] A book, published four years later, would herald the misadventure as *The Great Alaskan Dinosaur Adventure: A Real-Life Journey through the Frozen Past.*[42]

I first became acquainted with the expedition leader, John Whitmore, and his stated research goals when Robert King, the Bureau of Land Management's Alaska state archaeologist, contacted me to review Whitmore's fossil-collecting permit application. Whitmore's academic credentials

Fig. 3.7. No room on the ark cartoon. *Credit: John C. Holden.*

appeared adequate and his research goals, as stated, were quite acceptable. I could find no legitimate reason to recommend against the application. A little research into Whitmore's background and credentials did suggest that there might be a hidden agenda and that the participants might lack the kind of field experience that would support a safe and successful conclusion to the proposed expedition. However, my role was to judge the possible scientific merits of the proposed research and to make sure that any specimens collected by the participants would end up in the collections of the Earth Science Department of the University of Alaska Museum that I managed as curator. I did, however, communicate my suspicions as to the real goals of Whitmore and his team and my uneasiness with respect to their ability to complete their "adventure" on the Colville River and Alaska's North Slope. The permit was granted and the end result was to surprise even me.

The description I offer here of this bizarre episode in the study of Alaska's dinosaurs is based primarily on the published descriptions of the expedition in *The Great Alaskan Dinosaur Adventure* and is supplemented with accounts by residents of the settlement of Nuiqsut and communications between John Whitmore and Robert King and between the two of them and me. These communications extended over an eight-year period following the expedition. From the very beginning, as documented in the team's own words, the expedition was a debacle. They overloaded their inflatable rafts, which then deflated on hitting the water. They rented rafts and vital boat equipment sight unseen and then failed to inspect this equipment before launching into the icy Colville with only paddles for propulsion. They hadn't arranged any way to keep in touch with one another should they get separated, and not a single member of the party had any experience in the Arctic. If you set off in one of the more demanding and unforgiving places on the globe with so little expertise and preparation, you had better hope there is someone watching over you that will take special interest in you and your enterprise. It must be said that despite breaking most of the common sense rules of survival in the Arctic, this motley crew managed to collect some of the fossils targeted on their permit and reached the settlement of Nuiqsut, over 80 river miles (roughly 130 kilometers) downstream from Umiat, their Colville entry point. Although the book completely ignores over thirty years of previous geologic work and writing concerning this part of the Colville River and contains many inaccuracies concerning the dinosaur record and the Inupiaq village of Nuiqsut, it does contain a refreshingly candid account of the group's travails and near disasters. The book is also something of a contradiction. The back cover of the book refers to the area as "a remote, frozen wasteland," which is the image that the Arctic and Alaska in general conjure up for people who have never visited either. But how can this team of writers, who have been there, promote such an erroneous image of the Arctic tundra, the seemingly endless green carpet of plants (figure 3.3) that teems with animal life? How does anyone spend over a week on the largest North Slope river that is home to a wide variety of fish and a bewildering potpourri of birds, record the crossings of great masses of caribou, and still describe the environment as a "frozen wasteland?"

The lack of an appreciation for the true nature of the Arctic environment of Alaska was eventually matched by the lack of any significant results from the collections and subsequent research that this crew had indicated as possible outcomes on the permit application. No refereed science journal published the descriptions of their "unfossilized dinosaur bones" or the results of DNA analysis that was to be performed on frozen dinosaur bone in cooperation with a Russian colleague. As fate would have it, this quirky episode overlapped with the most scientifically productive period of my eighteen years of fieldwork and related research on Alaska's North Slope.

4 PEREGRINES, PERMAFROST, AND BONE BEDS

Digging Dinosaurs on the Colville River

The call of endless frantic days
Replaces the deep cold with
Permeating solar warmth
Life teems and searches all about
Springs forth to cache summer's riches
All of life caught up in frantic pace

The Challenges of Fieldwork in Arctic Alaska

One can make a case that field research anywhere presents unique challenges. However, having done geologic fieldwork in environments as diverse as the rainforests of southern Mexico and the deserts of California and Nevada, I can say that I find the Arctic of Alaska to be the most challenging of all. Perhaps it is the condensed time frame of the northern summer that results in the terribly frenetic schedule that all living creatures must follow or the weather that can change dramatically within a few hours from oppressive heat to wet cold or freezing. Perhaps it is the logistic demands of traveling over vast distances and covering terrain that offers few safe open spaces for fixed-wing landings and few navigable rivers. The rivers and large streams are particularly challenging because of the vagaries of summer storms that are spawned by the Arctic Ocean. The few large rivers are subject to dramatic, short-term changes in water levels along their courses. Water levels often change overnight, leaving boats high and dry, or worse, floating away downstream in lazy circles because they were not anchored securely. Beyond these physical challenges, I quickly learned about another environmental reality that could greatly affect access to localities and the timing of fieldwork along the Colville River: peregrine falcons.

The North Slope is a land filled with lakes and other wetlands. These aquatic havens and the long days of summer combine to produce a rich and welcoming land for feathered migrants from as far south as South America and as far west as central Eurasia. Individual lakes become nurseries for swans and other large birds. The shorelines of rivers and lakes become rich feeding grounds for shore birds, and the tundra thicket offers protection for nests, while the exploding insect populations feed swallows and their kin. The craggy bluffs adjacent to our field sites offer safe rookeries and hunting platforms for a wide variety of raptors, including peregrine falcons.

Peregrine decline worldwide by the late 1970s and early '80s elicited strict protection for these long-distance travelers, and this in turn, dictated when and where field activities were permitted. This typically took June and early July off the field calendar, and the weather usually deteriorates quickly after mid-August, leaving an extremely short field season. The expansion of protection to several other raptor species such as eagles would increase restrictions and shorten the field season even further by the late 1990s.

My introduction to fieldwork on Alaska's North Slope took place in July 1987, just a little over five months after I arrived in Fairbanks with a new wife and four bewildered cats to take up my position as associate professor of geology and geophysics and the curator of earth science at the University of Alaska Museum. I was no stranger to the challenges of geologic fieldwork in the high latitudes, since I had spent a summer on the Lena River in eastern Siberia and a month on the Yukon and Tatonduk rivers of Alaska in the late 1970s. I also had extensive experience organizing and directing complex field expeditions while teaching geology, paleontology, and environmental studies at Merritt College in northern California. However, I was now challenged to shift from invertebrate paleontological research to the unique demands of excavating, transporting, and studying dinosaur bones.

My Introduction to Fieldwork in the Arctic, 1987–91

I joined a team led by Professor William Clemens from my alma mater at the University of California Berkeley. This meant that I could ease into this new research challenge unburdened by the responsibility of logistics and detailed planning. I could concentrate on adapting to the working environment and sharpening my excavation techniques, as it was very likely that ultimately I would have to carry on the dinosaur research in Arctic Alaska. I had the good fortune to have the helping hand of the Bureau of Land Management's Northern District archaeological specialist, John Cook. He arranged for the perfect introduction to my future endeavors on the North Slope—a flight from Fairbanks to the landing site on the Colville River near Ocean Point in a Cessna 185—one of the workhorses of the "bush" that I would become dependent on in my future field research. The Cessna flight gave me a spectacular aerial traverse of the country that I needed to become fully familiar with in order to develop the proper geologic and geographic perspectives to do my job. I was also introduced to the ability of bush pilots to pack an amazing amount of gear into a small cargo space, a talent that would come in handy when preparing for future field expeditions on a limited budget.

Once you are airborne and heading north from Fairbanks, the seemingly endless patchwork of dark green forest interspersed with the lighter greens of willow and poplar is broken by the dark tannic brown of broadly meandering streams and stagnant thaw lakes that occasionally host a feeding moose or bear. Just as your senses become dulled by the sameness sweeping below, the sinuous silvery snake of the Trans-Alaska Pipeline leads the eye to the broad and deceptively smooth blue of the mighty Yukon River. As this watery immigrant from Canada fades from view, the aircraft's engine changes pitch as it gains altitude in anticipation of the massive snow-highlighted peaks of the Brooks Range that lie ahead. The brown and tan rock tones of these rugged mountains are broken only by glaciers or ephemeral snowfields nestled in gullies or rocky amphitheaters. The pilot guides the frail craft through a massive canyon, wingtips appearing to barely clear the jagged ridges and steep slide-covered barren slopes. For the geologist, this unfettered view of barren rock offers a kaleidoscopic view of the tremendous forces that record the ancient collision of crustal plates that left the rocks cleaved and bent into folds that a contortionist would envy.

All too soon, this spectacle gives way to a vast ocean of green, the Arctic Coastal Plain tundra. The pilot descends to 1000 feet and the tundra spreads out in sharp detail. At first, the landscape is broken only by an occasional brown-topped ice "blister," or pingo, and dark brown streams, but then you begin to come across the scars of abandoned river channels that are eventually joined by swarms of blue, bluish-green, and brown shallow lakes and nets of polygonal cracks.[1] Pairs of white swans and groups of ducks paddle lazily on these lakes fed by thawing permafrost. The pulse is quickened by the site of a huge brown bear sow and her cubs ambling along a shoreline. Some minutes later, the sight of a herd of caribou splashing as they cross a large stream induces awe. The Cessna banks hard left, and within a few moments, the massive tan bluffs and brownish-blue waters of the Colville River come into view on the horizon. Quickly the Cessna descends and positions itself over a large sand and cobble bar. The pilot flies low over the river edge of the bar, checking the winds and condition of the ephemeral landing strip. To the right, the tents of my colleagues appear like bright-colored mushrooms growing out of the sands and gravels of this emergent summer island. The plane banks hard left again just as it seems that we will crash against the bluff wall. The winds coming off the bluffs and the ground effect of the bar combine to lift and buffet the light craft and give it more gravity to absorb. Before you can fully adjust to this experience, the aircraft has landed and is noisily treading across the sand and cobbles on its huge tundra tires. After coming to a stop, the pilot revs the engine and turns toward the middle of the bar where a group of fellow campers await. Within minutes, there is a flurry of activity and the plane is quickly unloaded and, before you can catch your breath, the pilot is waving and heading away. This was an experience that my work on the North Slope would duplicate many times.

A large sandbar found near the northwestern shore of the Colville River and almost due west of Ocean Point (figure 3.1), became the main base camp and landing zone for most of my field seasons spent on the North Slope. The bar has acted as a landing strip and base camp for a series of dinosaur-collecting expeditions ever since the 1985 Clemens-Allison expedition, whose members informally named it the "Poverty Bar." There are a series of dinosaur-collecting sites along the beaches and bluffs across a narrow slough just northwest of the bar, which is located 0.6 miles (nearly 1 kilometer) upriver from the Liscomb Bone Bed. The Liscomb Bone Bed and closely related rocks extend along the base of massive 90- to 100-foot (28- to 31-meter) bluffs on the north and western bank of the lower stretch

Fig. 4.1. Alaska Flyers Cessna bush plane at Poverty Bar base camp, Colville River in 1992. *Credit: Roland Gangloff.*

of the Colville River. Named for Andrew W. Colville in 1837 by British explorers, this largest of the North Slope rivers has been given various names by the resident Inupiaq natives, including Kupik and Gubik.[2] Flowing 350 miles (over 500 kilometers) from its headwaters on the north slope of the De Long Mountains, a western extension of the Brooks Range, it empties into the Beaufort Sea after forming a large, complex delta west of Prudhoe Bay. This greatest of North Slope rivers would be my "highway" for future exploration and fieldwork on the North Slope.

The Liscomb Bone Bed: Challenges and Triumphs

The Liscomb Bone Bed is assigned to the Prince Creek Formation and is part of a sequence of sedimentary and volcanically derived rocks that were deposited by air fall and large meandering rivers at, or substantially above, the present latitude of 70° N.[3] This organic-rich, mudstone-dominated bone bed is, at the most, a yard thick. However, the remarkably dense accumulation of dinosaur bones and teeth in it earn it a label as a legitimate bone bed, equivalent to any described in the paleontological literature.[4] The position of the bone bed at the base of high steep bluffs makes its excavation and study especially daunting.

The bone bed is Upper Cretaceous in age (between sixty-eight and seventy-three million years ago). This age range estimate is based on the potassium-argon and argon-argon isotopic dating of volcanic ash beds and specific mineral crystals found in rocks above, below, and between a series of separate bone beds that include the Liscomb Bone Bed near Ocean Point on the Colville River.[5] This would place the Liscomb Bone Bed close to the boundary between the Campanian and Maastrichtian stages

Fig. 4.2. A. View of Liscomb Bone Bed at high water in 1989; the bone bed is marked by black lines. B. Teachers mapping in quadrat on Liscomb Bone Bed, 1992. *Credit: Roland Gangloff, Tony Fiorillo, and David Smith.*

of the Late Cretaceous. Fossil plants and invertebrates, in concert with biostratigraphic methods, place the Liscomb Bone Bed in the Early to Late Maastrichtian and therefore within the critical last five million years of the Cretaceous period. This interval of time was when the much heralded Cretaceous-Tertiary (now the Cretaceous-Paleogene) boundary and the concomitant great series of extinctions occurred that eliminated the nonavian dinosaurs as well as from 65–70 percent of the species that were living at the time.[6] The sequence of sedimentary rock that includes the Liscomb Bone Bed is overlain by uncemented and poorly consolidated marine and nonmarine sediments primarily of Pleistocene age (1.81 million years ago). The entire North Slope is underlain by hundreds of feet of perennially frozen ground called permafrost. The Pleistocene sediments are not fully frozen like the older rocks beneath and thus contain discontinuous lenses or wedges of ice that partially melt enough to form thaw lakes on the tundra surface on the top of the bluffs. By July, when the North Slope has reached its thermal maximum, the ice wedges produce enough meltwater to destabilize the younger unconsolidated sediments that then form debris and mud flows (see plate 4). At the same time, the older rock below thaws along exposed surfaces to as much as 3 to 5 feet (1.5 meters) into the bluff. This all makes for a dangerous unstable situation at the base of the bluffs, where, of course, the bone bed must be excavated and mapped. In addition, the distance into the bluff that can be worked is greatly limited by the depth of permafrost thaw. At times, the gullies along the bluff face contain accumulated snow from the previous winter that adds to the meltwater volume above the bone bed.

As I joined my colleagues from Berkeley on the Liscomb Bone Bed that first summer, I was also introduced to another challenge as an excavator—Alaskan mosquitoes. They are not particularly large, as many would have you believe, but they can be very abundant and persistent. This is definitely a factor that interferes with concentration and clear thinking when working on the bone bed. The mosquitoes are particularly distracting and irksome when you encounter them for the first time, but some folks never seem to adjust to their bothersome presence, and they leave never to return. The use of DEET-dominated repellents, the most effective type, comes with nasty side effects. I found that I needed to develop a zenlike approach in which I almost totally absorbed myself in my work to put these insects at mental bay. Despite what many newcomers to the Arctic think, mosquitoes did not evolve to harass man and beast. Indeed, they are the principal pollinators in the Arctic.

The field season of 1989 proved to be atypically warm and fraught with abundant rock slides that disrupted our work. Two colleagues and former costudents at Berkeley, Thomas Rich and Patricia Vickers-Rich, came to visit and were appalled at the dangers and challenges that the Colville bluffs presented. Thom and Pat were so struck by their experience on the North Slope in the summer of 1989 that they vowed to work with me to find a better and safer way to work the bone beds and other dinosaur-rich deposits along the Colville. Their concern for our field dangers was all the more impressive since they had to literally mine dinosaur remains on high cliffs along Australia's rugged coast, south of Melbourne.[7] Their anxiety and cooperation would lead to a unique approach to dinosaur excavation on the North Slope some twenty years later.

Following the initial collecting and evaluation of the Liscomb Bone Bed and related strata for their fossil content that took place from 1985 to 1990, it became my responsibility to pursue the detailed mapping, collecting, and taphonomic analysis of the Liscomb and related beds. Taphonomy is a branch of paleontology that focuses on how once-living organisms become fossils—in other words, it examines the processes of death and fossilization. Taphonomic studies require the collection of fossil specimens and the detailed mapping of those specimens in three dimensions. Today, this is greatly aided through the use of high-resolution GPS devices and laser-ranging instruments, but in the 1980s and early '90s, such high-resolution electronic aids were not available.

The methodologies adapted to the Colville bluffs and outcrops consisted of shoveling and clearing away talus, digging into the bluff for 3 to 6 feet (roughly 1 to 2 meters), if the permafrost would allow it, and setting up quarry squares (quadrats) that measured 1 × 1 yard (roughly 1 square meter). Once a quadrat area was cleared and the slope was terraced to make conditions safer, steel rebar posts were driven into the ground to delineate each square. Driving the posts through the rock and remnant permafrost while keeping them level and upright usually required a great deal of effort. Wire, which was strung using a level, connected the posts on three sides. This provided a reference system used to measure horizontal distance from the quadrat's sides, as well as vertical distance (depth), measured from the top of the bone bed using a line attached to a plumb bob (see figure 4.2B), which allowed for mapping, in three dimensions, the location of each specimen as it was collected. However, mapping and collecting were done while kneeling or lying on cold ground and at the base of steep bluff walls. Neophytes quickly learned why foam knee and sleeping pads were highly recommended. The Liscomb Bone Bed presented another challenge to excavating teams. The bone bed is vertically graded, that is, the upper two-thirds are dominated by small bones, teeth, and bone fragments. Generally, the skeletal elements become larger the deeper you go. The largest and most complete bones are found at the bottom of the bed. This meant that a great deal of time and effort were spent in mapping all of the small bones,

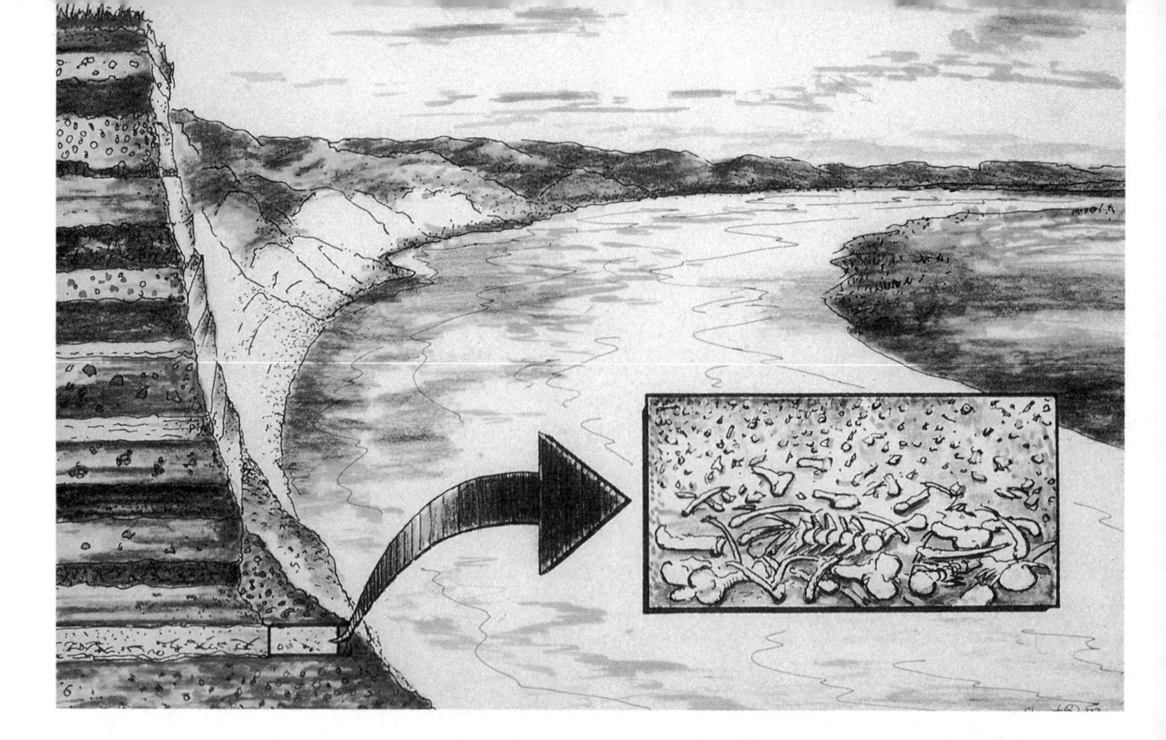

Fig. 4.3. Cartoon illustrating the stratigraphy of the Colville bluffs and details of the vertical distribution of skeletal elements in the Liscomb Bone Bed. *Credit: Tom Stewart and University of Alaska Museum of the North, Fairbanks.*

teeth, and fragments before the really "cool" bones were encountered (see figure 4.3).[8] Mapping and cataloging these earlier encountered specimens demanded a great deal of focus and discipline. Earlier excavations had pretty much ignored the contents of the upper part of the bone bed in a "race" to get to the larger and more complete skeletal elements. During some field seasons, very few excavations reached the bottom of the bone bed due to inclement weather and/or landslides, but I was determined to document the entire bed.

Quadrats were worked by teams of two to three persons, depending upon the size of the field group that could be recruited each field season. This allowed each member of the team to exchange the grueling mapping for a period of data recording and note taking. Data recording was made more mentally demanding by swarming mosquitoes and drippy wet weather. The team approach also had the benefit of making the working environment safer, since the data recorder was charged with keeping an eye out for slides or cascading debris from tunneling ground squirrels. A number of us were convinced that the squirrels, or "sik siks" as the Inupiaq natives called them, were deliberately tossing the debris down on us. I found that Arctic ground squirrels were much more of a danger and nuisance than any other mammal, including bears. In my eighteen years of fieldwork in Alaska, I only had to deal with an Alaskan brown bear once. For safety and legal reasons we always carried rifles while in the field. However, I never had to fire a single shot at a bear in defense or for any other reason. The best defense against bears is to be with a noisy group and to keep a clean camp with food stored in bear-resistant containers. However, Arctic ground squirrels (*Spermophilus parryii*) can be a great deal of trouble in camp and at fossil sites. Not only do these ubiquitous rodents produce underground galleries along the edge of high

Fig. 4.4. Arctic ground squirrel in ramen box at Poverty Bar base camp, 1996. *Credit: Roland Gangloff.*

cliffs that can collapse under you, but they chew through tents and "steal" food. Even though they are primarily seed- and plant-eating rodents, I found that they would climb on our meat-drying racks and abscond with drying caribou meat. They are awfully cute until you step into one of their ankle-breaking holes or get hit by debris falling down from their burrows at the top of high bluffs.

Excavation of dinosaur bones and teeth was accomplished with a set of simple tools—knives, awls, ice picks, chisels, geology hammers, trenching shovels, and an array of brushes. The entombing rock parted rather readily, aiding the freeing of bones and teeth.

However, the parting also cleaved larger bones, which were often highly fractured as a result of yearly freezes and thaws and the compressive load of the enclosing sediments. In addition to being vulnerable to fracturing, the larger bones were also subject to a degradation that turned some of them partially into a dark-brown powder; this degradation may have been related to the amount of the mineral pyrite that was contained in the bones along with silica and calcite (see plate 3 and figure 4.5C).[9] As a result, the bones demanded the addition of large volumes of various consolidants and adhesives to strengthen and protect them. Even though all of the bones and teeth were mineralized, many of the larger or more fragile bones and teeth had to be encased in plaster jackets before they could be safely extracted. The extra support and protection that these jackets afforded also aided in getting the skeletal parts back to the museum in good shape, and in most cases, with no new fractures to repair.

Along with these challenging conditions, the Liscomb Bone Bed held one other surprise that I hadn't dealt with before. The bone bed was very rich in finely dispersed carbonized plant remains and had a total organic content that made it oily and very dark. The organic content was high enough to make a petroleum geologist or geochemist salivate over its potential for petroleum storage if found in the right structural context.

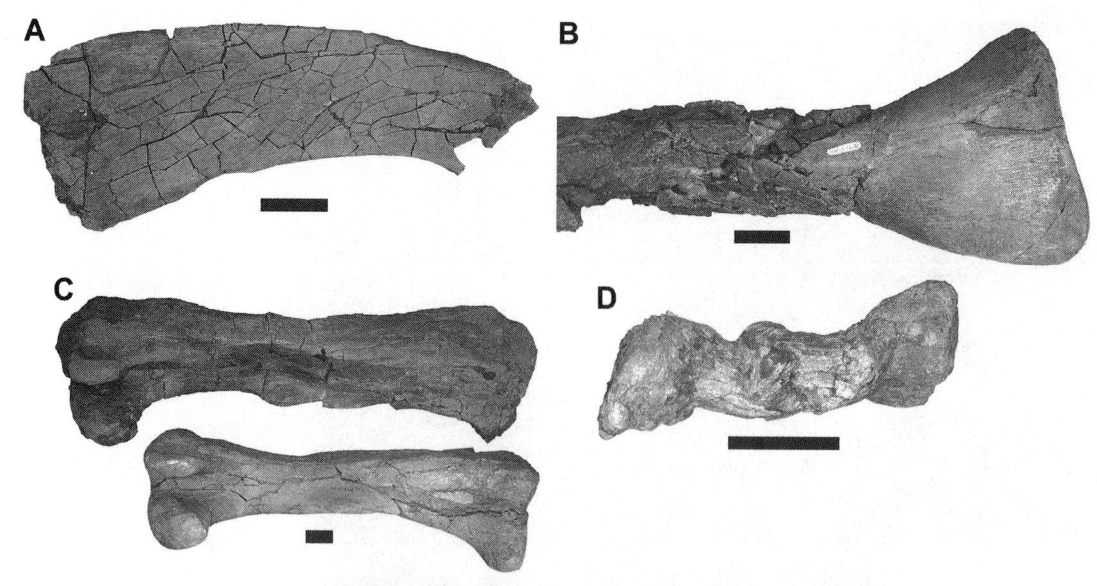

Fig. 4.5. Bone modification in Liscomb Bone Bed. A. Mosaic fracture pattern in partial left scapula. B. Dense transverse and oblique fracturing of diaphysis of left tibia. C. Pair of left femora that exhibit typical range of preservation. D. Densely fractured and plastically distorted metacarpal. Scale = 4 centimeters. *Credit: Roland Gangloff and Tony Fiorillo.*

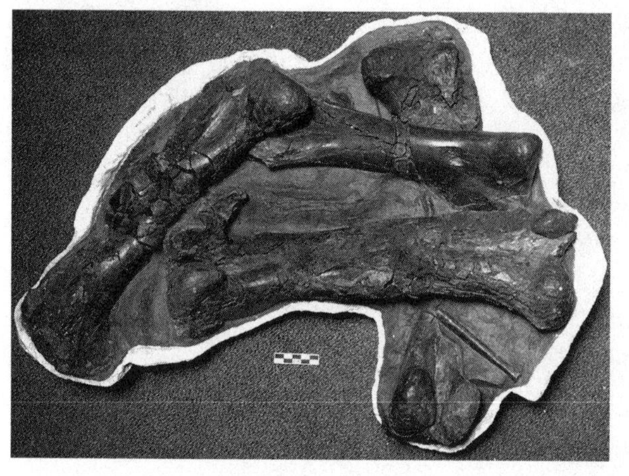

Fig. 4.6. Bone "jam" from bottom of Liscomb Bone Bed. Includes two left and one right femora, a partial right tibia, and a partial dorsal rib of *Edmontosaurus.* Scale = 5 cm. *Credit: Photo by Roland Gangloff; preparation of bones by Mark Goodwin and Kevin May.*

Unfortunately, for fossils excavators this attribute just made note taking and cleanup all the more difficult.

The first seven years of field collecting and study of the Liscomb and associated sedimentary sequences clearly pointed to the remarkable fossil abundance and interesting taphonomy encountered along this stretch of the Colville River. By the end of 1993, the first major field expedition totally under the auspices of the University of Alaska Museum had been completed. This and previous expeditions resulted in nearly a thousand identifiable bones, teeth, and fragments being collected and curated in Fairbanks and the Museum of Paleontology at the University of California Berkeley. It was becoming increasingly apparent that the Alaskan Arctic might become one of North America's dinosaur treasure troves. It was also becoming clear that the Liscomb and closely related bone beds were greatly dominated by small individuals. That these small individuals were juveniles rather than dwarfs was supported by microscopic osteological study and gross anatomical measurements and comparisons with published data.[10] In addition, the bones and teeth

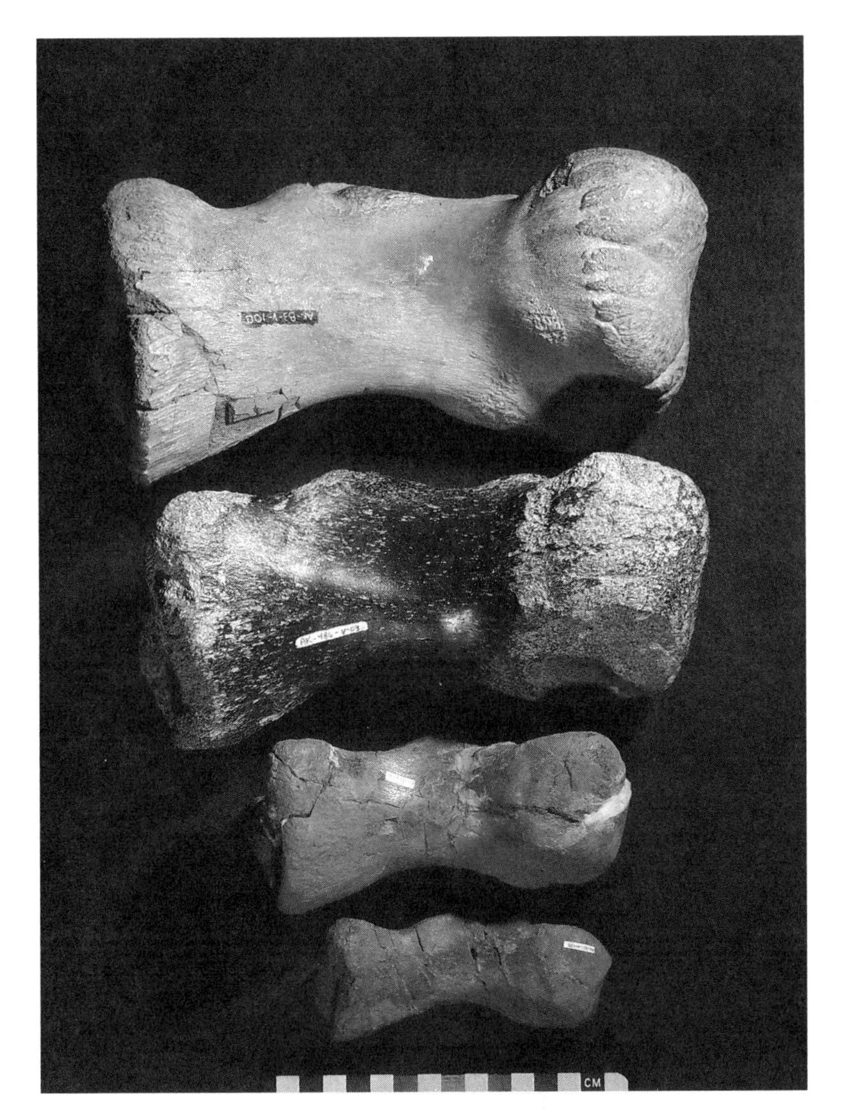

Fig. 4.7. Array of *Edmontosaurus* right metatarsal IV foot bones showing range of size (subadult at top to late juvenile at bottom) and range of color (permineralization) from Liscomb and related bone beds. *Credit: Roland Gangloff.*

from the Liscomb Bone Bed were also compared to collections at the Museum of Paleontology and Royal Tyrrell Museum of Palaeontology. These studies were conducted by me and several students. The most detailed and thorough micro-osteological study was done by a graduate student, Wendy Ehnert who had participated as a high school teacher in the 1993 field expedition.

They Seem to Be Everywhere: More Discoveries Along the Colville River and South-Central Alaska, 1993–99

Even though the Liscomb Bone Bed and closely related beds remained the principle focus of research, over the next seven years our knowledge of the geographic breadth and geologic depth of the dinosaur record in the Arctic greatly expanded. At the end of this period, the Liscomb and two new bone beds yielded close to four thousand specimens attributable to at least ten different dinosaur genera representing five major evolutionary lineages. Five of the genera were theropods, carnivore-scavengers. The years

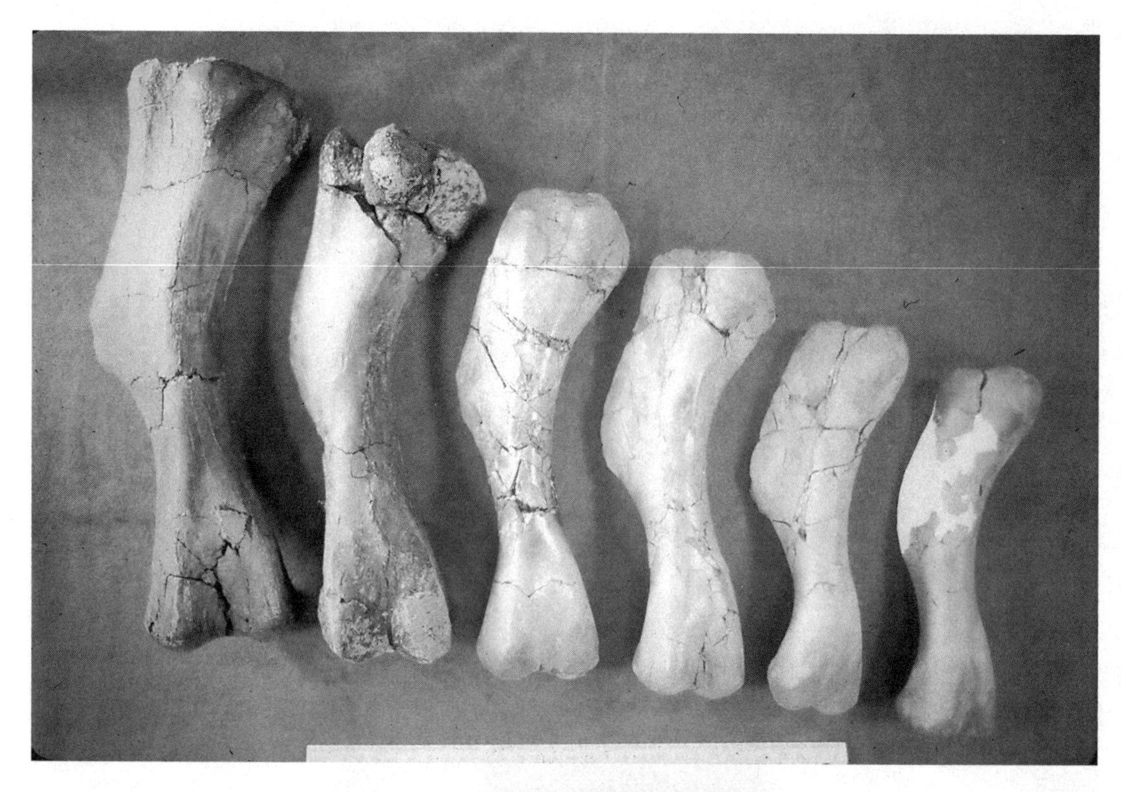

Fig. 4.8. Array of juvenile *Edmontosaurus* left humeri from Liscomb Bone Bed showing the range of sizes and morphology. Largest belongs to a very late juvenile, the smallest belongs to an early juvenile. Scale = 30 centimeters. *Credit: Roland Gangloff.*

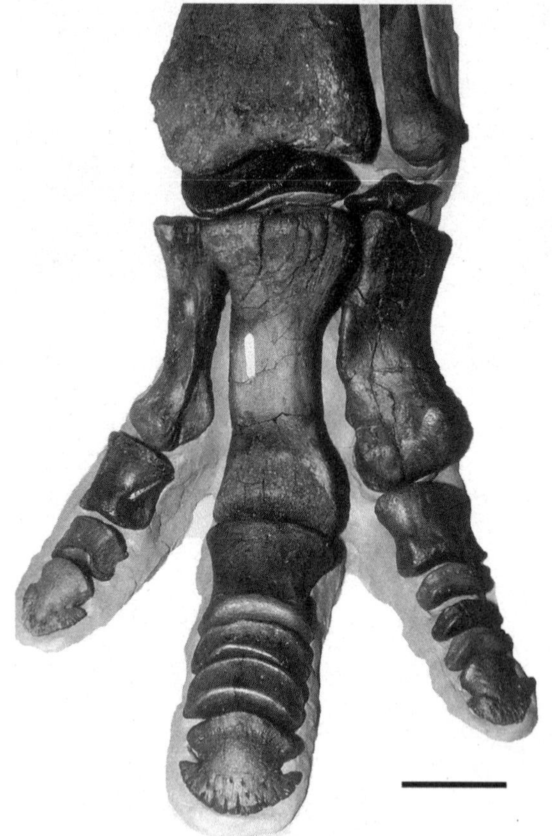

Fig. 4.9. Composite left foot and distal end of hind leg of a juvenile *Edmontosaurus*, Liscomb Bone Bed. Scale = 5 cm *Credit: Gary Grassi and Roland Gangloff.*

1994, 1996, and 1999 were especially productive as field crews worked the Liscomb and adjacent bone beds. These years also witnessed larger combined numbers of volunteers, students, and teachers than in previous years. Field parties went from six to more than twenty participants per expedition. In addition, the years 1997 and 1998 witnessed teams from both the Fairbanks and Anchorage campuses of the University of Alaska, working from rafts, discovering twelve new trackway sites of both dinosaurs and birds.[11] "Trackway" as used here refers to a sequence of tracks or footprints made by dinosaurs or birds that represent one or more individuals. The footprints were originally made in wet sand or mud that later turned into rock (figure 2.4, A_2, and C_2). These abundant footprints were part of a thick sequence of older Cretaceous rocks along the upper to middle section of the Colville River, between the confluences with the Awuna and Killik rivers (see figure 3.1).[12]

Anchorage Colleagues Join the Effort

The cooperative effort between the two University of Alaska campuses began in 1994. It was in the winter of this year that my colleague at the Anchorage campus, Professor Anne Pasch, contacted me concerning a recent discovery near Anchorage. A hunting guide from Wasilla had found a poorly preserved skull in the central Talkeetna Mountains just 93 miles northwest of Anchorage and gave it to Anne for study. This exciting find proved to be the first fossil of an armored dinosaur, or ankylosaur to be found in Alaska or the Arctic. More specifically it was a member of the family Nodosauridae and attributable to the genus *Edmontonia*.[13] This -find in Alaska significantly increased the latitudinal distribution of ankylosaurs in North America. Soon after, a student of Anne's discovered an early type of duck-billed dinosaur while he and his family were out on a berry-picking trip. These two finds added to the Cretaceous record of dinosaurs in marine burial grounds.[14] In addition, the Talkeetna hadrosaur turned out to be one of the earliest record of this clade in North America.[15] Dinosaurs suddenly seemed to be popping up all over Alaska. Prior to this, a few scattered footprints of dinosaurs had been reported from two sites on the North Slope and one from Jurassic-age rocks on the Alaskan Peninsula (see figure 2.1).[16] Anne's experience with the Talkeetna discoveries got her excited about continuing our cooperative research on Alaska's dinosaurs. Anne asked me where I thought our combined efforts might generate the most exciting and significant results. We directed our attention to the upper and middle section of the Colville River. Our research of the literature turned up several finds that suggested that there was a high potential for new dinosaur discoveries on this part of the Colville. The rocks were mapped as representing an earlier part of the Cretaceous than those that contained the Liscomb Bone Bed farther down river. Gil Mull had published a record of a single dinosaur footprint at a site on the Colville, near the confluence with Awuna Creek (see figure 3.1).[17] If you find one footprint, there are bound to be more.

The Banner Years of 1996, 1997, and 1998

The field season of 1996 was particularly significant since it was marked by the longest season ever (early July to mid-August) and by the participation of the largest number of volunteers and staff. Two groups totaling twenty-one members worked on the Liscomb Bone Bed during two separate periods. The efforts of these two teams produced nearly one thousand specimens and their related maps and notes. For the first time, mapped quadrats were completed near the northern end of the Liscomb Bone Bed. This field season also produced the only hadrosaur sacrum ever collected from the Liscomb. It was the most complete articulated skeletal unit that has thus far been collected from this bone bed.[18] Three short, articulated segments of hadrosaur tails have since been collected, but no articulated skeletal units, such as skull, pelvis, or limbs, have been found as yet. Disarticulation of the skeletons is a hallmark of the Liscomb and stratigraphically related bone beds thus far. The general lack of articulation of skeletal parts such as the skull and limbs reflects the lack of closure and fusion of sutures. When this observation is combined with measurements of the bones, and then compared to published and unpublished datasets for *Edmontosaurus*, they support the conclusion that the Liscomb Bone Bed is dominated by juvenile *Edmontosaurus*.[19]

Fig. 4.10. R. Gangloff and upright fossil tree trunk in flat-lying sediments at Lili Creek. *Credit: Robert King, Bureau of Land Management, Anchorage.*

The years 1997 and 1998 hosted two banner field seasons. In 1997, Anne Pasch and I prepared to investigate the upper part of the Colville River near Awuna Creek. We had patched together a series of small grants from our respective University of Alaska campuses and logistic assistance from the Bureau of Land Management. Having accomplished this, we then recruited volunteers who could pay for their trip and began making detailed plans to raft down the Colville River to the logistic base at Umiat. Unlike most North Slope expeditions that I had led over the years, this one began at the general aviation airport, where seven of us boarded a chartered aircraft that took us to the airstrip and facilities at Umiat (see plate 5). From there we were ferried by small fixed-wing aircraft to our first field camp around 60 air miles up the Colville River. We planned to visit localities that had been noted as containing fossils by oil company geologists and members of the U.S. Geological Survey and Alaska's Division of Geological and Geophysical Surveys. We would prospect for evidences of dinosaurs and other fossil vertebrates between the recorded localities.

The plan was to meet a crew of four volunteers led by Dave Norton that was boating up from Nuiqsut some 80 river miles (130 kilometers) to the south of Umiat. The upper Colville part of the plan went very well. I was picked up by a Bureau of Land Management helicopter with Robert King on board. Robert is the bureau's chief Alaska archaeologist and is responsible for overseeing all fossil collecting work on bureau-administered lands in Alaska. Robert and I were flown to a locality with upright, in place, petrified trees that a team of ARCO oil exploration geologists had found the year before. Within an hour of our landing, Robert and I had found several trees that had been buried quickly by a series of heavy influxes of mud during floods. In addition to these upright trees, we found abundant slabs and sections of petrified trees littering the small stream valley below the standing trees. We must have looked like two children in "candyland"

Fig. 4.11. Joanne Avery with tyrannosaurid tooth, Kikak-Tegoseak beach, 1997. *Credit: Roland Gangloff.*

to the helicopter pilot as we rushed about hooping and hollering to one another as we came across more and more pieces of petrified trees. Robert and I took photos, specimens, and notes before our allotted hour was up—helicopter time being very expensive. But this was not to be the only "fossil rush" for the day.

Robert and I were soon on our way by copter to a site that Gil Mull, a geologist for the state of Alaska, had found and reported in a paper for a scientific journal.[20] This side trip turned out to be even more exciting and significant than the one to the petrified forest. Again, we were given only an hour to survey the site. However, Robert and I located Gil Mull's original site and found the first trackway, close to where Mull had reportedly found a single footprint. I returned to the base camp flushed with excitement and with my mind racing with the possibilities of even more significant finds in the near future. I leaped from the helicopter and ran over to Anne Pasch and the assembled volunteers and reported our good fortune. Anne and I immediately began to plan a trip for 1998 that would return us to the Mull site for more documentation and prospecting at what appeared to be a very promising location.

Meanwhile, the Norton-led crew that was boating up the Colville River ran into difficulties due to low water. They were forced to turn around before they reached our rendezvous point at Umiat, greatly disappointed, not to mention totally exhausted by their efforts. However, they were able to add a couple of important data points to our record of dinosaurs on the North Slope. The most important was the discovery of a bone bed that contained the bones and teeth of an unusual type of rare horned dinosaur that is confined to northern North America, *Pachyrhinosaurus*. This discovery was coupled with a beautifully preserved tooth of a tyrannosaurid theropod where none had been found before.

Fig. 4.12. Kevin May holding a natural cast of a *Tetrapodosaurus* footprint from Nanushuk Formation, Killik bend, Colville River, North Slope, Alaska, 1997. *Credit: Roland Gangloff.*

While the Norton-led team was struggling up the Colville, Anne and I led our team down the river. A site near the confluence of the Colville and Killik rivers known as the Killik bend (see figure 1.2) yielded trackways and several individual prints that quickly expanded the known record of these trace fossils.[21] These important dinosaur traces at Killik bend were joined by several abundant fossil plant localities that were documented as we drifted and paddled our way down river to Umiat. By the time we reached Umiat, it had become clear that we would need to organize and fund another rafting trip that would allow for more prospecting and collecting on this stretch of the Colville River. July turned out to hold hot, dry, sunny weather. This was why the river was so low and why our Norton-led

team had such a frustrating trip—weather on the North Slope is always a challenge no matter what the season. Just two years later, the Poverty Bar base camp was flooded out in the middle of our field season for the first time since the base camp had been established.

Arctic Rivers, Always a Challenge

Arctic rivers are notoriously difficult to navigate. Depths and water levels are constantly changing due to weather and the high influx of sediment load that the rivers must carry. If the main channel becomes too shallow due to sediment fill or you end up on a smaller channel by accident, you can find yourself engaged in the grueling exercise of hauling boats and rafts over and around partially submerged sandbars for a mile (2 kilometers) or more. The smaller rivers often become highly braided and shallow during the summer. On the North Slope, heavy rains may pound the slopes of the Brooks Range while the weather is clear locally. Winter freezing can form thick ice dams in the upper parts of tributaries than can hold back the initial surges of rainwater. When these ice dams break, the cascade of water travels tens of miles down channels and surges into larger river channels with startling results. Being so close to the Arctic Ocean, the North Slope often sees violent rain squalls during the summer that can produce hurricane velocity winds. Being in a raft or small boat during such a wind storm can be life threatening, as winds whip the water into 3- to 4-foot-high chop and the cold rain or sleet reduces vision beyond a few feet. Going overboard into 38–40° F water will sap your body heat in minutes. This is no environment for those who can't appreciate these physical dangers and the rapidity with which Arctic weather can change. Navigating these challenging rivers is essential in order to do any detailed geologic mapping or prospecting; it is simply impractical to use helicopters to do the kind of detailed fieldwork that required for sound paleontological research.

1998 Holds Surprises and Surpasses the Successes of 1997

The 1998 field season was supposed to begin with a two-day dusty bumpy drive up the 400 plus miles (650 kilometers) of gravel known as the Dalton Highway. The Dalton Highway was a vital link to the North Slope for all of my field seasons. It allowed us to drive within 70 to 100 miles (112 to 160 kilometers) of our main field camps. From there, bush planes could be contracted to cover the final distance. This was a much less expensive way to reach our destination compared to chartering flights from Fairbanks to Deadhorse (Prudhoe Bay) or Umiat.

Our two vehicles left Fairbanks with a combined team of three principal investigators and four students from the Fairbanks and Anchorage campuses of the University of Alaska. As usual, the group left Fairbanks and headed north on the Elliott Highway in a cheerful and excited state. The Elliott would eventually take us to the much longer Dalton Highway (locally called the "haul road" because it follows and services the Trans-Alaska Pipeline [see figure 1.2]). But within a few miles, we were turned back toward Fairbanks by Alaska state troopers. A sniper had fired on a vehicle, and all traffic was blocked for the foreseeable future. As both vehicles were equipped with four-wheel drive, the decision was made to try an alternate route that would circumvent the closed stretch of the highway. Two hours and 10 miles later, the very dusty and highly "shook-up" group was finally on its way again to the Dalton Highway, and hopefully, in time to meet the Alyeska Pipeline Service Company's helicopter at pump station no. 4. The helicopter was being provided to fly in some of our bulky equipment. This assistance was only available n that day and was vital to keeping within our transportation budget.

We traversed the Brooks Range through the steep and spectacular Atigun Pass and arrived at the pump station a bit later than the appointed time. We lucked out—the pilot was late as well and willing to change plans. We then took advantage of the luxury of the long summer days and joined the pilot in the station dining room and were treated to a prime rib dinner with all the fixings accompanied by a blizzard of questions from the station staff about our plans. These Trans-Alaska Pipeline stations are the workhorses that pump oil from the complex of wells and service facilities at Prudhoe Bay to Valdez some 800 miles to the south (see figure 1.2). These stations often present an almost surreal picture, surrounded as they are, by rugged wilderness. Some, like pump station no. 4, sit atop mesa-like prominences that call to mind the ancient castles of Europe that dominated their domains.

Several times over the years, these pump stations and their crews hosted me, my colleagues, and my volunteers, providing much appreciated warm welcomes, hot food, and showers. The Alyeska pipeline stations are not open to the public, but the company and its crews had been generous with help over the years, and it was a great comfort to know that they would be available 24/7 to assist us in an emergency.

After sending half of our equipment with the Alyeska helicopter, we left pump station no. 4 and proceeded north on the Dalton Highway to a small airstrip at Happy Valley near the Sagavanirktok River (see figure 3.1). This airstrip hosted the bush pilot and aircraft that then flew us to our first campsite on the Colville near the confluence with Awuna Creek. Even with the assistance of the helicopter from Alyeska, it still took three flights to transport the seven members of our team and our gear, which included three inflatable boats and motors. Finally, we were all together on the southern shore of the Colville River, a few miles east of the mouth of the Awuna. We needed to assemble our inflatable rafts, check the outboard motors and other critical equipment and supplies, set up our tents, and get some much-needed sleep. It was 1:00 AM and still very light by the time most of us retired.

By the next afternoon, the team had finally settled on a place to make camp. We wanted to be near the center of the half-mile long exposures that we were calling the "Mull" site. The sighting of clear, deeply impressed three-toed footprints 100 feet from the river on our arrival portended what was to become a dinosaur trackway bonanza. Four days camped at this location yielded four other locations within a half mile (one kilometer) that held well-preserved trackways made by at least two other dinosaur types. Two held four-toed footprints that were most probably made by

Fig. 4.13. *Tetrapodosaurus* footprint with skin impression at Mull trackway site, Nanushuk Formation. *Credit: Barry McWayne and University of Alaska Museum of the North, Fairbanks.*

Fig. 4.14. Dinoturbated trackway surface at Mull site, Nanushuk Formation. Scale = 17 centimeters. *Credit: Roland Gangloff.*

Fig. 4.15. Lizzie May, Yupik Alaska Native, pointing to side view of dinosaur footprint, Nanushuk Formation. Near confluence of Awuna and Colville rivers, 1998. *Credit: Kevin May.*

horned or armored dinosaurs.[22] These tracks also held impressions of skin patterns. At one location, the trackways were so numerous and dense that individual tracks and trackways were hard to delineate—an example of "dinoturbation" of the original muddy surface.[23]

The team finally set course downriver with hundreds of photos, an 8-foot-long latex peel (mold) of one of the trackways, a 6-foot tracing of a short theropod trackway, and a collection of rock plates with individual footprints. Most importantly, the group was heading downriver after four long and physically exhausting days with high morale. The quality and diversity of the dinosaur tracks and trackways that they had discovered and documented was outstanding. Ten more significant dinosaur track and trackway sites were discovered and carefully documented with photos, notes, and GPS readings. A few representative specimens were collected when possible. These vertebrate ichnite samples were joined by a variety of invertebrate body fossils and traces. A wealth of fossil plant specimens would also join these other specimens. Of particular importance was the revisiting of the 1997 site of Killik bend. The imprinted rocks are spectacularly exposed at high angles of inclination, challenging the mind and body.[24]

As we continued downstream, we came across a stark reminder of the dangers of conducting fieldwork on the North Slope, the gnarled and twisted wreckage of a Cessna 180 that had crashed while attempting to a make a landing on an untested cobble and sand bar. The pilot was one of my students who had been a volunteer on the '97 expedition and had

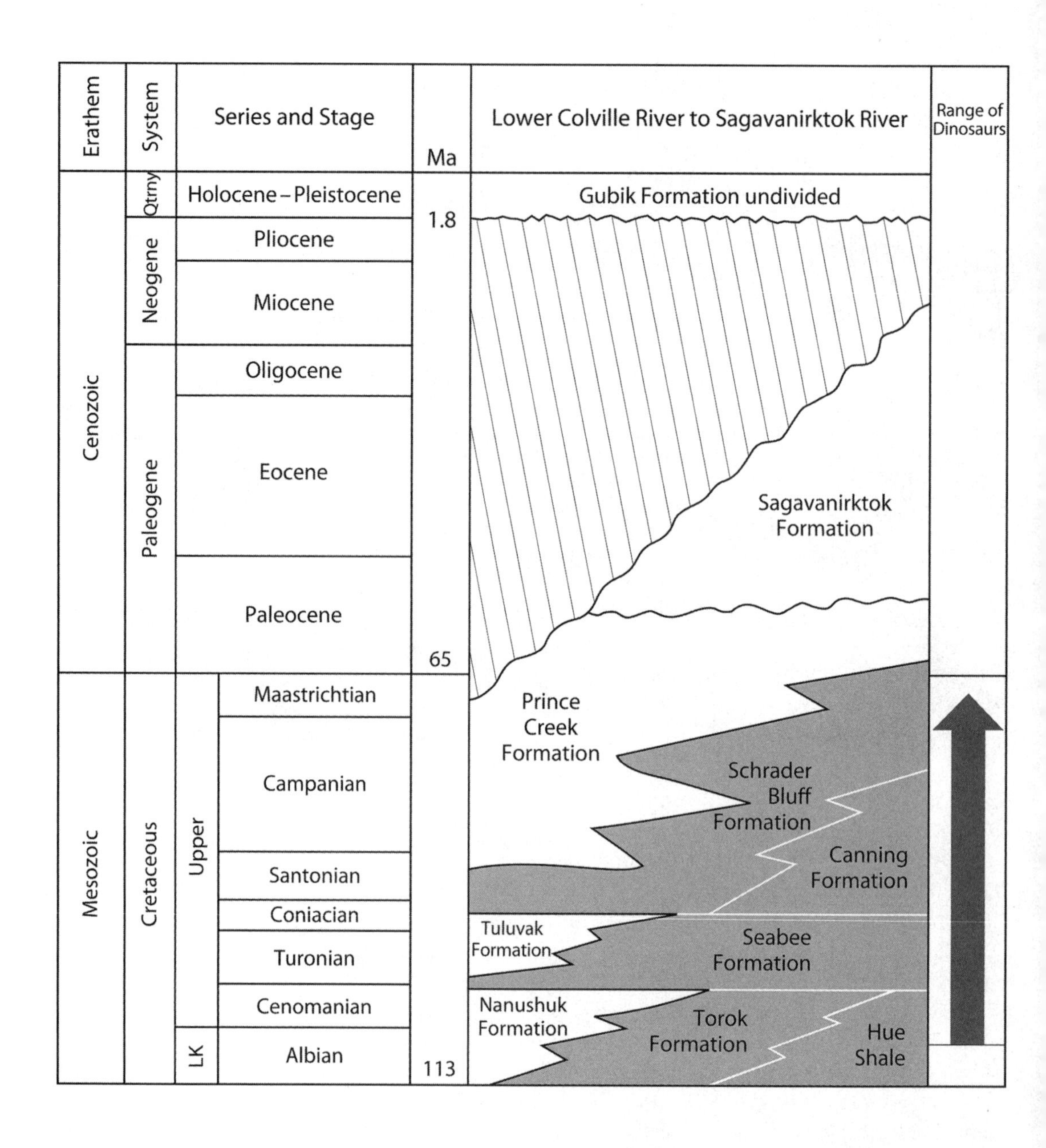

Fig. 4.16. Chronostratigraphic column for lower Colville and Sagavanirktok rivers with range zone of dinosaurs. Formations in gray are predominantly marine. *Credit: Dixon Jones (modified from Mull, Houseknecht, and Bird 2003 and Gangloff and Fiorillo 2010).*

decided to do some "poking" around on his own following the rest of the group's departure from the field. He survived, but the wreckage brought home once again just how difficult bush flying was.

The next two days were taken up with more successful prospecting. Three sites held partial trackways or isolated tracks, and the fossil floral record was expanded. It was becoming quite clear that the Nanushuk Formation held a rich assemblage of ichnofossils other than dinosaur ichnites. The variety of ichnites and floral remains deserved much more effort than we were giving them. Future expeditions were certainly warranted. Kevin May and his daughter Lizzie also discovered several new sites in the younger Prince Creek Formation. We finally arrived at our take-out point—the facilities at Umiat. Before we all went our separate ways, we spent two days repacking, finishing field notes, and enjoying our time relaxing and listening to O. J. Smith recount his early flying experiences, which we countered with tales of our river adventures.[25]

Umiat—Only in Alaska

Umiat is one of the strangest places you will ever visit. Presently it is an unincorporated settlement with no permanent residents and has a modern history that reaches back to 1940, when the U.S. Navy drilled a test well close to the present location. Umiat became an important naval base in 1944, when the navy established the Naval Petroleum Reserve no. 4 and made Umiat the logistic hub for oil and gas exploration on the North Slope. The navy built an airstrip that could accommodate multiengine aircraft and set up a fueling center and accommodations for staff. By 1952, eleven test wells had been drilled, and the U.S. Geological Survey had compiled geologic field data that would act as a starting point for the great North Slope oil rush of the early 1960s. In late July 1998, Umiat was a collection of old vehicles, rusting trailers, and an array of dilapidated buildings of various sizes. The best of the buildings had been erected by Alaska's Department of Transportation to house crews and equipment to maintain the almost 6,000 feet (1,800 meters) of gravel runway (see plate 5) that could easily accommodate C-130 Hercules transports. The "herc" had been the main aerial workhorse during the development of the Prudhoe Bay oil fields and the subsequent Trans-Alaska Pipeline. The frenetic oil exploration days now over, this airstrip and the related facilities has turned into a more low-key logistical base that hosts all manner of bush aircraft, helicopters, and the occasional chartered "short hop" airliner or "herc." Umiat also serves a critical function as an emergency airstrip and refuge from bad weather or bad planning by general aviators flying north across the Brooks Range.

In 1997–98, the self-proclaimed mayor of Umiat, O. J. Smith, ran the fueling station, weather station, and classic bush "inn." Most importantly, he provided hot showers and good food. If you were seeking entertainment, you would find it in O.J., who never ran out of stories about his North Slope and Brooks Range adventures flying his Piper Super Cub. He especially liked to recount his experiences with cheechakos (tenderfoots) who would stumble into his territory. Smith took over the operations at Umiat in 1975 with his two sons, providing bush flights for hunters and fishermen throughout the summer until his death in 2001—the passing of one of the more colorful characters that ever occupied the North Slope.[26]

Umiat has now become a logistical center for research parties from several universities as well as state and federal agencies, including the Bureau of Land Management's Northern District Arctic Field Office. This means that there are all manner of studies being conducted from or that are reliant on the facilities at Umiat.[27] Global warming is being documented in this part of the Arctic, and Umiat is a critical ground-based weather station that greatly aids weather data gathering and weather prediction in northern Alaska. Bird, caribou, moose, and musk-ox studies are being conducted, and their researchers are crossing paths with geological, paleontological, and archaeological field parties. Umiat and the North Slope fall within the North Slope Borough's political jurisdiction and are subject to its management and consultation. By the 1990s, our dinosaur research projects required three federal and three state permits, as well as a North Slope Borough permit. This added a new complexity to fieldwork planning that soaked up a great deal of precious time in and out of the field.

Why Did So Many Dinosaur Tracks Go Unreported for So Long?

Before leaving Umiat, I sat down with my co-leaders and students and tried to explain how it could have happened that we had found such an extensive record of dinosaur tracks and trackways in a stretch of the Colville River that had been traversed by tens, if not hundreds of federal, state, and oil company geologists over some forty years. We were puzzled by what appeared to be a major oversight by scientists who in other ways had done such laudable geologic studies and had published extensively on them. Gil Mull was a notable exception since he had found and published on one track site. In addition, a single preliminary U.S. Geological Survey report had illustrated fossil bird tracks from the Colville River. We all had a hard time understanding why such a rich fossil record had been either overlooked or unreported. After a lot of interchange and speculation, we came to the conclusion that most of the geologic studies were focused on mineral and petroleum resources, and that the time spent on the ground between helicopter flights seldom allowed for the kind of observations that are needed in order to clearly see bird tracks and dinosaur trackways. It is important to survey potential sites while experiencing low-angle sunlight, and that often requires being on a particular part of a rock outcrop for a number of hours. It is important to view the rock surfaces as the sun goes through its complete overhead transit. It also requires the "dumb luck" of arriving at a track site just when the sun is at the best angle to reveal the impressions or natural casts. In addition, vertebrate fossils are usually not considered important data for the evaluation of mineral and petroleum potential. Properly documenting them also requires special, time-consuming techniques. It should be noted that invertebrate fossils such as clams were collected and consistently reported in geologic publications by the same geologists who didn't include any references to vertebrate fossils.

The results of the 1993 to 1999 field seasons greatly expanded the known dinosaur record for Alaska and the Arctic despite a two-year hiatus in excavation and mapping of fossils. The entire period witnessed the addition of nearly twenty-five hundred bones and teeth. Even more important was the discovery of a new bone-bed site packed full of the remains of, *Pachyrhinosaurus*, a distinctly northern ceratopsian dinosaur.[28] This bone bed is the only site in the Arctic that contains an accumulation of horned dinosaur remains. In addition, preliminary measurements and evaluation of the thousands of *Edmontosaurus* bones and teeth pointed clearly to the abundance of juvenile individuals that, in turn, indicated that some dinosaurs overwintered in the Arctic. When all of these discoveries and the extensive skeletal dataset were taken into account, it became obvious that the record of Alaska's dinosaurs had much to tell us about dinosaur paleoecology as well as about postulated Cretaceous dispersal routes and faunal exchange between North America and Eurasia. These same finds began to bring into focus Alaska's faunal connections to Alberta, Canada, and perhaps even Baja, California. By now, I was beginning to hear an echo of Thom Rich's 1994 prophetic pronouncement that I might be sitting on one of the most fossiliferous Cretaceous dinosaur areas in the world, perhaps an area that was even as abundantly endowed with Cretaceous-age dinosaur remains as the world-famous Red Deer River valley of Alberta, Canada.

Contributions of Teachers, Volunteers, Students, and Research Colleagues

Working with K–12 teachers from Alaska and the lower forty-eight states was a wonderful and enlightening experience. They proved to be enthusiastic, tenacious, and highly conscientious even under strenuous and challenging conditions. They learned mapping and excavation techniques rather quickly and were grateful to have a role in a real science project. Most had only learned their science from textbooks and prepared lab exercises, and some had no formal science training at all. Some, much to my surprise, had never slept on the ground or camped in a small tent. We had some great conversations about how science and scientists really work, as well as more specific discussions about dinosaurs and why Arctic finds were demanding that we rethink our views of dinosaur ecology and extinction.

The teachers impressed me with their willingness to put up with swarms of mosquitoes, rapid changes in weather, and long days. A number had come to the North Slope fearing that they would be attacked by bears and learned that our most constant wildlife nemesis was actually the Arctic ground squirrel. One male teacher had finally had it with these rodents and their antics while eating lunch. He picked up a geology hammer that lay near him and flung it at a particularly tenacious "furball" that was attempting to get into a food cache. He had intended to scare the critter away, but he nailed it with the sharp end of the hammer and pinned it to the ground. He had to retrieve the bloody hammer and dispose of the body. I will never forget the look on his face; he was truly saddened by the episode. Luckily, Gary Selinger, one of my closest colleagues from our museum, was able to ameliorate some of the teacher's mental pain with his usual good humor and practical approach to life. Gary was an experienced archaeological excavator and became a great instructor and "cheerleader." He was our unofficial morale booster, constantly going from quadrat to quadrat encouraging individuals, offering snacks, and heading off angst and frustration. Every successful field program is usually endowed with some version of Gary Selinger.

Before I leave my recounting of experiences with teachers in the field, I must mention a wonderfully delightful person by the name of Barbara Gorman. Barbara was a librarian at one of the elementary schools in Fairbanks and knew most of the teachers that worked with us. Barbara had filled a slot that had been vacated by a teacher who had withdrawn just before we were to leave for the North Slope. Barbara, at age sixty-eight, turned out to be the oldest member of her group, and she eventually participated in the greatest number of field expeditions to the North Slope of any of my volunteers—starting in her late sixties and ending in her late seventies. Most

Fig. 4.17. Quadrats with warming stove on Liscomb Bone Bed. April Crosby and Merritt Helfferich in foreground with Herm Seibert and Roland Gangloff beyond, 1996. *Credit: Bill Hopkins.*

importantly, she proved to be an inspiration to everyone with her ready smile and great enthusiasm for fossils, fieldwork, and the out of doors, no matter what the weather or conditions. Barbara also became a wonderful advocate for the dinosaur research program by bringing the results of our work into the classroom with her "dino box." This modified suitcase was packed with teaching materials, including specimens that I loaned to her. Barbara and the dino box traveled to many parts of Alaska and to venues like Texas and Germany where she visited family and friends. Barbara is still going strong on her "crusade" to spread the word about the dinosaurs of Alaska and the Arctic.

An interesting and diverse cadre of highly dedicated and effective volunteers added to the accomplishments of the teachers. Three stand out most in my mind. The first was a bear of a man named Kevin May. Kevin is a squat, physically powerful man, whose demeanor and rugged looks tend to mask a bright and highly perceptive mind (see figure 4.12). He is the kind of person that is most dependable when things get the roughest, and he quickly grasps the larger picture. He has a ton of field experience in Alaska and loves to be out in the "bush." Kevin ultimately became a highly valuable assistant at the museum and a coauthor of several technical publications on Arctic and Alaskan dinosaurs. Kevin is still on the staff of the University of Alaska Museum of the North in Fairbanks and continues to take part in research on the North Slope.

Bill Hopkins and Herm Seibert were a dynamic duo and a delightfully matched pair. They came my way after first being volunteers for Thom Rich and Patricia Vickers-Rich at Dinosaur Cove in Victoria, Australia. They came highly recommended and proved to be a valuable asset to our quarry work as well as to the general operation and morale of our base camp. They were a constant source of good cheer, and they delighted our crews with their pithy tales of life experiences and accounts of other fossil digs. They both were highly energized and childlike in their enthusiasm over finding fossils. Herm was a retired engineer, and Bill had been executive director of the Alaska Oil and Gas Association in Anchorage, Alaska. No job was beneath them, and their sense of humor made even privy duty or trash disposal a smile-producing event. I shall always remember the bright but devilish glint in their eyes and the enthusiasm with which they approached tasks such as developing a set trap for ground squirrels near our supply tent.

I never thought that I would be directing lawyers, engineers, teachers, and high-placed executives to dig latrines, shovel out quarry pits, and work for hours on their hands and knees for the opportunity to unearth the deep past. It is always regenerative to find that anyone, no matter their age, still has a child within them that is just dying to be released. We had some great discussions around the campfire as to the implications and value of our daily discoveries. After all, good science requires that we recall our youthful curiosity and inquisitiveness and shed the blinders of false sophistication and narrow perception that adulthood tends to foster.

My closest and most supportive colleague during my years of work on the North Slope was Dave Norton. His contributions to the dinosaur

program are too numerous to list all of them here. First and foremost, Dave took on the bulk of the logistical burden of our fieldwork. This meant procuring, operating, and maintaining, our principal river boats. First, he used his own boat and motors and then he obtained a larger boat from the U.S. Bureau of Mines and a brand new 90 HP Honda engine that he funded with a grant from the Pew Trust. Dave recruited students from Barrow and constantly sang the praises of our program. He garnered logistical and other support from several North Slope Borough agencies, including facilities to house overwintering supplies and equipment. Dave was a constant source of energy and encouragement, and he helped me to overcome many of the frustrations and barriers that fieldwork in Arctic Alaska often generates. I have no doubt that his presence and support kept me from giving up several times during my tenure in Alaska.

Recognition of Our Efforts Brings Needed Support

Several grants obtained during the 1990s provided significant increases in monetary and logistical support for dinosaur research on the North Slope.[29] The expanding recognition that came from television, newspaper, and magazine coverage of our work in Alaska helped to recruit volunteers as well as to convince university, local, and federal officials of the importance of the new finds. In addition to increased governmental support, greater publicity also helped to garner valuable logistical support from ARCO Alaska and the Alyeska Pipeline Service Company, which runs and maintains the Trans-Alaska Pipeline for a consortium of oil companies.

Several interviews on local public television informed a segment of the general public that mammoths and mastodonts were not the only large vertebrate fossils in Alaska and that dinosaurs had, in fact, lived, thrived, and died in the ancient Arctic of Alaska as well. Colleagues throughout the world began to take more notice with the presentation of research results at professional meetings and the publication of the results of our research in professional journals. In 1997, a new research program, the Arctic Alaska Dinosaur Program was defined and heralded in the journal *Arctic Research of the United States.*[30] In the same year, a chapter in the acclaimed *Encyclopedia of Dinosaurs* was devoted to the rapidly expanding record of dinosaurs in the polar regions.[31] National and international interest was generated by two television documentaries. Although these activities were often a drain on my time and patience, in and out of the field, they also opened several doors that had been closed or unknown to me.

As I look back, I recall the wonderful group of colleagues and volunteers that made so much of my fieldwork and research possible in the face of small budgets. I realize that over one hundred individuals donated some seven thousand hours, or nine hundred-plus person days to the enterprise. This allowed me to accrue the most extensive and diverse record of dinosaurs in either polar region. I hope that my publications do justice to their trust and hard work.

TEXAS, TEACHERS, AND CHINOOKS

Taking Fieldwork to a New Level

Over 3,000 miles (4,800 kilometers) of rugged mountains and high plains separate Alaska's North Slope and Dallas, Texas, today. Most people would not think of Texas and Alaska as being close in any way. But there is quite a history that brings Alaska and Texas much closer than their current geography would ever suggest. In terms of "deep time," they were once connected by a great Cretaceous age interior seaway and long food-rich shorelines that supported abundant populations of duck-billed and horned dinosaurs. In recent times, petroleum technology and economics have closely bound Alaska to Texas and other so-called "oil patch" states.[1]

The Texas connection doesn't stop at the two paleontologists at the Vertebrate Paleontology Laboratory, Texas Memorial Museum in Austin, Texas, who were the first to study and describe the dinosaur bones collected by Robert Liscomb. Another colleague from Texas, Anthony (Tony) Fiorillo from the Museum of Science and Nature in Dallas, first joined me during the field season of 1998, and we have been working together on the record of Alaska's Arctic dinosaurs ever since. Tony was responsible for involving me in paleontological surveys for the National Park Service at Denali National Park and Preserve and Katmai National Park and Preserve. These programs ultimately laid the foundation for discoveries of the first evidence of dinosaurs in Denali National Park and at Aniakchak National Monument (see figure 2.1).[2]

Tony had extensive experience with vertebrate taphonomy and a history of working with the National Park Service. This combination broadened and strengthened the research on dinosaurs in Alaska—something that was needed, since I was one of only two vertebrate paleontologists employed full-time in the entire state of Alaska. By 2005, a close-working relationship with Tony had resulted in the securing of several small grants, including one from the Jurassic Foundation, and a large National Science Foundation grant, as well as several coauthored journal papers and presentations at professional meetings.

California Teachers, the University of California Museum of Paleontology, and the Arctic Alaska Dinosaur Program

The years 2001 to 2005 witnessed a remarkable coming together of support, resources, and recognition for the Arctic Alaska Dinosaur Program.[3] This period would also include my retirement from the University of Alaska. The main event of this remarkable period, was the repartnering between the University of California Museum of Paleontology and the University of Alaska Museum in the dinosaur program. This resumption of a partnership between the two institutions added a new aspect to the relationship. The partnership was based on combining the very successful teacher outreach program at the University of California Museum of Paleontology with the field research on the North Slope that had defined the Arctic Alaska Dinosaur Program . Judy Scotchmoor, who was director of outreach and public programs at Berkeley had been a volunteer for the dinosaur program in 1999 and enthusiastically proposed that a group of teachers from California be brought up to Alaska as participants in a geoscience program. Funding was secured through a Geoscience Education grant from the National Science Foundation, and nine teachers were selected to spend a month during the summer of 2002 in Alaska. The teachers received instruction and firsthand experience in geology and paleontology during a traverse from Anchorage to Fairbanks and then, after further instruction at the University of Alaska, Fairbanks, proceeded up the Dalton Highway to Prudhoe Bay and the Deadhorse Airport (see figure 1.2). The teachers then flew by bush plane from Deadhorse to the Poverty Bar base camp on the Colville River. Another exciting new dimension to the field program was that it would receive the logistical support of the U.S. Army in Alaska.

Chinooks and the Army Deliver the Goods

Dave Norton and I worked closely with the commanding officer, Major Lissa Young, and her staff of B Company, 4th Battalion, 123rd Aviation Regiment, U.S. Army, Alaska. during the winter and spring of 2001–2002. The plan was to have the company's four heavy-lift CH-47D Chinook helicopters and crews support two of our field camps with transporting field crews and gear to and from the field. In addition to the food, shelters, camp gear, and the personal gear of the expedition members, Dave, the army loadmasters, and I were to plan for the delivery of our 18-foot river boat, fuel, and engine by one of the Chinooks—not an easy task for even this famous aviation workhorse. The aviation crews of Company B, known as "Sugar Bears North," were stationed at Ft. Wainwright in Fairbanks, relatively close to my museum at the University of Alaska. The Sugar Bears were to combine a training mission to the North Slope and Arctic Ocean with the provision of logistical support for two separate field camps on the Colville River. It had taken a year of hard work on the part of Major Young to shepherd the mission through the army's chain of command. The mission represented an unprecedented commitment to a teaching and research program by the U.S. Army.

The Chinooks and their highly skilled aviators and load crews transported our large river boat and most of the gear and supplies needed to sustain the teacher-volunteers and research staff at the Liscomb Bone Bed base camp for two weeks. The delivery of the large aluminum riverboat grabbed the rapt attention of the camp crew, who were watching from the shore of the base camp river bar. The Chinook pilot hovered just above the swirling waters of the Colville River while the load crew lowered the rear ramp, and as the pilot brought the aircraft to a 20° angle, the boat was "birthed" onto the river by two crew members on board, as bystanders, on shore, stood by in awe and admiration—unfortunately, no one thought to take a photo. This maneuver was good practice for special operations that the crew might be called upon to perform in a battle zone. The Sugar Bears then flew Tony Fiorillo, Louis Jacobs, and a crew comprised of graduate students and a museum preparator onto a river bar some 30 miles (48 kilometers) upriver across from a site designated as Kikak-Tegoseak. This site was named in honor of the Inupiaq grandparents of Ron Mancil, the Alaskan native student that discovered the bone bed in 1997.[4]

The Teachers Arrive at Poverty Bar

While Dave Norton, Kevin May, and four crew members from the University of Alaska Museum set up the base camp at Poverty Bar, elementary through high school teachers from the West Contra Costa schools in the San Francisco Bay Area made their way up the Haul Road. Several stops and short hikes were scheduled along the way to introduce the teachers to the spectacular geology exposed along their route. Two of the highlights along this trip were wading in the Yukon River and posing at the Arctic Circle marker, a new experience for most of the teachers and a welcome "photo-op."

Eventually, the group reached the North Slope's main airport at Deadhorse—not the most comforting name for anyone who was about to experience their first bush flight into a remote part of Alaska's Arctic.[5] Deadhorse is the largest airport on the North Slope and primarily serves the petroleum complexes of British Petroleum and Conoco-Phillips on the shore of Prudhoe Bay. The short summers witness the influx of thousands of tourists who arrive by bus and plane from all over the world. They come to see the abundant birds, caribou, brown bears, and, on occasion, the odd polar bear as well as to dip a toe in the Arctic Ocean-Beaufort Sea. Deadhorse Airport is the center for bush flights for hunting, scientific research, and ecotourism as well as a scheduled stop for Alaska Airlines and Frontier Flying Service. The airport was developed on a caribou summer feeding ground. Caribou are given the right-of-way at all times. Several times while leaving or returning from expeditions I have experienced a "wave off" from the control tower while caribou were blocking runways.

The nine teacher participants flew onto Poverty Bar with some of their personal gear and stashes of granola bars and other hidden snacks—insurance against starvation if they crashed on the tundra—along with me, Judy Scotchmoor, and two volunteers who helped drive and manage camps along the trip up the Dalton Highway. Awaiting this group of Arctic "cheechakos" was Dave, Kevin, Kevin's daughter Lizzie, and two experienced staff from the University of Alaska Museum, Amanda Hanson and Barbara Gorman. This field party of eighteen was one of the largest groups ever to participate in the mapping and excavation of the Liscomb Bone

Bed. The teachers were organized into teams of two or three and by the next morning were engaged in the hard work of clearing out the accumulated talus and setting up their quadrats (quarry squares) along the midsection of the bone bed.

After a few days of adjusting to the mosquitoes, becoming familiar with the mapping routine and instruments, the teacher teams settled down to the day-to-day routine of digging, measuring, drawing, and recovering the specimens of bones and teeth from the cold ground. Typically, small pieces of bone were uncovered, along with individual teeth and the ubiquitous ossified tendons that serve as strengthening cross-members in the lower spinal architecture of hadrosaurs. These smaller skeletal specimens elicited squeals of joy and much excitement during the first few days, but after having to map and remove so many tiny pieces, team members began responding to the routine with comments such as "Oh! Not another ossified tendon."

Mapping required manual dexterity and concentration. Since measurements were taken in three-dimensional space, lines had to be held level for accurate horizontal measurements and a plumb bob had to be used to establish a true vertical distance below the top of the bone bed. The vertical measurement required the most concentration and dexterity. The direction or geographic orientation of the larger and longer bones necessitated using a Brunton compass to take a bearing adjusted to "true" or geographic north.[6] The same instrument was used to measure the dip or inclination of the bone in reference to the horizontal plane. This measurement was needed to determine if the bone was lying flat or not, a question that relates to variables such as current velocities and other hydrodynamic factors during the deposition of the bone. Certain patterns can point to the bone having been disturbed subsequent to its original position after death. The measurements were accompanied by a scaled sketch of each specimen as it related to the reference boundaries of the quadrat. Teeth always received very spirited responses no matter where they were encountered and especially if they were from one of the four theropod taxa in the bone bed.

As the days passed, the teams began to unearth more and more recognizable bones until by the end of the ten days, they were coming across fairly large and mostly intact bones—the exciting treasure that rewarded the patience and tenacity required to reach the bottom of the bone bed. Overall, the teachers did a superb job with all aspects of the mapping process and quickly became proficient with the consolidation and gluing and subsequent plaster jacketing of the bones. This field season produced over 1,000 specimens with the mapped quadrats accounting for 892 of these.

The Return to Fairbanks: The Teachers Learn an Important Lesson

As most of the teachers and support crew at Poverty Bar began to slow down and finally take a rest after hours of intense activity connected with breaking down and packing the base camp, the unmistakable sound of the huge overlapping rotors of the Chinook could be heard in the distance. Before long, the first of three Chinooks was forming a swirling dust devil as it landed about 100 yards (98 meters) away. You are always mentally ready

Fig. 5.1. Teacher Rena Cutright in quadrat on Liscomb Bone Bed with plaster-jacketed leg bone of *Edmontosaurus,* 2002. *Credit: Phelana Pang and Christopher Tolentino.*

Fig. 5.2. Teachers assembled at the Chinook and raring to go after ten days on North Slope, 2002. *Credit: Phelana Pang and Christopher Tolentino.*

to leave once you have taken your tent down, so it is always exhilarating to see that first aircraft that assures you that you will not be left behind. It is not a fully rational feeling, but even though I relied on aircraft for extraction out of the field for some thirty-two years, I found that the feeling never fully left me. As soon as the Chinooks powered down completely, the teachers ran to greet their deliverers.

Within an hour, the cavernous hulls of the three CH-47Ds were filled with all the gear and the dinosaur diggers themselves and were lifting off and heading south. Following the Colville River, the Chinooks soon flew just east of the Kikak-Tegoseak site, where another Chinook was extracting the Fiorillo-Jacob's team of excavators. In a little over an hour, the Endicott Mountains and the spectacular Anaktuvuk Pass came into our view. The flight followed along the north fork of the Koyukuk River to the remote community of Bettles and its airstrip. The view of all of these landscapes from the large open ramp of a Chinook is quite breathtaking and most of the teachers sat mesmerized by the panorama.

Fig. 5.3. Looking north through the rear ramp of a Chinook to the Brooks Range and Anaktuvuk Pass and River below. *Credit: Roland Gangloff.*

The weather and the group's spirits could not have been better. The Chinooks began a long descent to Bettles, a village with thirty to forty permanent residents, a large airstrip, and a strategic position near the Gates of the Arctic National Park. Bettles serves as an important fueling stop for the jet-powered, fuel-thirsty Chinooks on such a long trip. Within an hour, we were lifting off once again. Continuing south, we crossed the Dalton Highway and its companion pipeline and flew over the Yukon River bridge before landing on the playing fields of the University of Alaska in Fairbanks.

On returning to Fairbanks, the teachers participated in the essential process of preparing and readying their specimens for placement in the Earth Science Department's collections. Proper curation required that each specimen be assigned and labeled with a unique alphanumeric code. Once this was done, descriptions and field context data were entered into a computer database. This entire process can be tedious and even boring to some, but it is absolutely critical to making the specimens valuable for subsequent analysis and to using them as the basis for hypotheses as to how they became the fossils that were collected and as to their role in the history of the Earth.

Before the teachers left Fairbanks for home, they caught up on sleep, curated several hundreds of the eight hundred-plus specimens that they had collected, and enjoyed some of the local attractions, hokey as those might be. The teachers would soon face the new school year with renewed enthusiasm and lots of great tales.

Once I had returned to Fairbanks, relieved of the heavy responsibilities of conducting field operations, I was able to reflect on the whole undertaking and put it into perspective. We had been very fortunate. The weather had been exceptionally good over the entire time the teachers were in Alaska. There were a few cold and misty days but no heavy storms or wind. There were no major injuries or deadly accidents. Very little equipment had been lost or badly damaged. We now had the luxury of satellite radios and could keep in touch with our counterparts upriver and with emergency services in Barrow or Prudhoe Bay, adding a much-needed safety feature that Canadian field researchers in the Arctic had had the benefit of since 1959. This was a vast improvement over the unreliable UHF/VHF handheld radio system that marked the first twelve years of my research on the "Slope." Our only backup in those years were personal

Fig. 5.4. Teachers unwrapping and cataloging fossil specimens at the University of Alaska Museum that they collected on the Colville River. *Credit: Roland Gangloff.*

locator beacons that were monitored by the North Slope Borough in Barrow over 160 miles (266 kilometers) to the northwest.

Kikak-Tegoseak and Other Field Programs in Alaska

The Kikak-Tegoseak locality had been an important part of my field program on the North Slope ever since its discovery in 1997. Fieldwork in 1998 and 1999 with my colleagues, Dave Norton, Kevin May, and Tony Fiorillo confirmed the presence of a rich pachyrhinosaur bone bed that also held evidence of three other dinosaur taxa—a hadrosaur, ornithomimid, and tyrannosaurid. The discovery of the Kikak-Tegoseak pachyrhinosaur bone bed was preceded, in 1988, by the sharp-eyed Howard Hutchison who discovered the first pachyrhinosaur in Alaska. Howard Hutchison, of the University of California Museum of Paleontology, came across the end of the snout of a skull projecting from a sandstone bed as he was prospecting along the base of high river bluffs just upstream from the southern end of Poverty Bar. Howard soon had excavated most of what turned out to be the main body of the cranium with only part of the frill intact. The lower jaw was missing but the distinctive thick nasal boss was hard to miss (see figure 5.8). The skull was an isolated specimen contained in a channel sandstone unit stratigraphically below an *Edmontosaurus*-rich bone bed found a hundred feet (nearly 30 meters) to the northeast, and 0.6 miles (0.9 kilometers) southwest of the Liscomb Bone Bed. The Kikak-Tegoseak bone bed is approximately 30 river miles (50 kilometers) south

Fig. 5.5. Graduate student James Sammons with volunteers during excavation and mapping of bone bed near top of the bluff at the Kikak-Tegoseak site, 2001. *Credit: Roland Gangloff.*

Fig. 5.6. Lizzie May and Linda Casassa proudly show a plaster jacket containing a *Pachyrhinosaurus* skull found at the Kikak-Tegoseak Bone Bed in 1998. *Credit: Kevin May.*

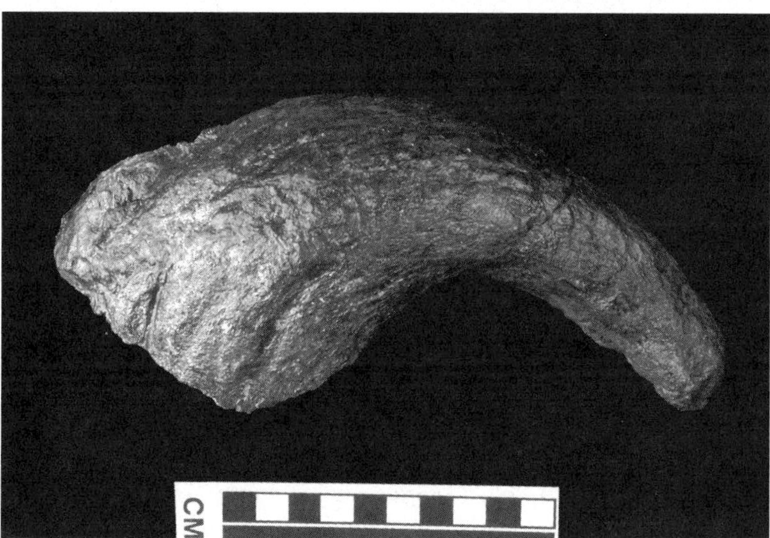

Fig. 5.7. Cornuate spike or process from the parietal frill of *Pachyrhinosaurus* found at the Kekak-Tegoseak Bone Bed. Scale = centimeters. *Credit: Barry McWayne and University of Alaska Museum of the North, Fairbanks.*

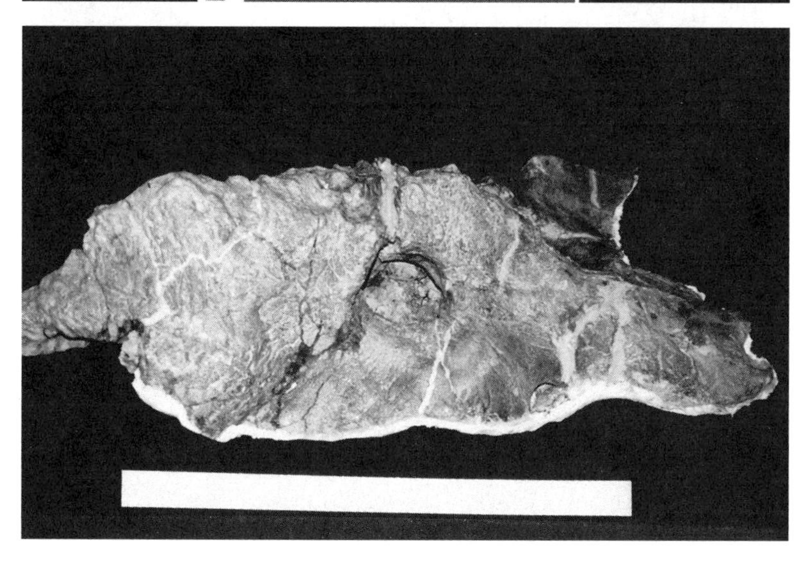

Fig. 5.8. Partial skull of first *Pachyrhinosaurus* collected in 1988 on the North Slope, 30 miles downriver (north) from Kikak-Tegoseak Bone Bed. Scale = 1 meter. *Credit: Gayle Nelms and University of California Museum of Paleontology, Berkeley.*

of where Howard found the first pachyrhinosaur, and is stratigraphically below and therefore somewhat older.

Kikak-Tegoseak Dinosaurs Challenge the Chinook and U.S. Army

The heavy-lift capability of the CH-47D Chinook Helicopter at 28,000 pounds (12,600 kilograms) and the ingenuity of its crew were sorely tested at the Kikak-Tegoseak site at the end of the 2002 field season. The army's rationale for supplying four of these expensive aircraft to our research programs was the valuable experience in Arctic and mountain flying that would be gained by its crews. In addition, the commanding officer wanted the crews to face logistic challenges in the Arctic environment. A great variety of navigational and logistic chores and challenges were connected to the establishment of the base camps at Poverty Bar and Kikak-Tegoseak. However, the Chinook assigned to take the Fiorillo-Jacobs team back to Fairbanks at the end of the field season, "ran" into a particularly stubborn piece of the past. A group of three *Pachyrhinosaurus* skulls, still partially embedded in rock and enclosed in a large plaster jacket, took much more work and skill than had been anticipated.

The large jacket and skulls sat on a jumbled slope near the top of a 115-foot- (35-meter-) high bluff. The large helicopter had to take a precarious position above the jacketed skulls in order to keep the distance between the rear ramp and the skulls short enough to lessen the strain on the cable and winch. Unfortunately, the angle of repose taken by the helicopter combined with the weight (approximately 1,000 pounds) of the jacket and skulls put too much strain on the helicopter's cargo bay winch. The steel wire snapped, jamming the winch and forcing the crew to abandon the precious cargo. A second attempt, two days later, ended in success and the "Sugar Bears" being able to maintain their pride.[7]

2005: A New Team Takes on the Challenge

The summer field season of 2005 was my last on the North Slope. It was also the first of a three-year National Science Foundation–supported program that would combine the efforts of the Gangloff-Fiorillo team and a team put together by Paul McCarthy. Paul was a faculty-research colleague from the Department of Geology and Geophysics and the Geophysical Institute at the University of Alaska, Fairbanks. Paul brought his expertise in terrestrial sequence stratigraphy and sedimentology to bear on working out the fine details of the depositional environments represented by the bone beds and the packages of sedimentary and volcanic rock that they were part of.[8] He was aided by four graduate students. This coalition of museum paleontologists and sedimentary geologists would represent the second such University of Alaska-led multidisciplinary team to attack the detailed history of the Late Cretaceous-age Prince Creek Formation and its rich larder of dinosaur fossils. The field season of 2005 not only increased our understanding of the detailed history of the physical sedimentary record of the Prince Creek Formation but also added several new dinosaur-bearing localities. In addition, a very rare skeletal component, a partial braincase, of the theropod *Troodon formosus* was collected.[9] This specimen joined another that was collected earlier from a quadrat in the Liscomb Bone Bed. Together the two specimens confirmed the presence of this fascinating large-brained and large-eyed member of the lineage that gave rise to birds. Previously, this taxon had been reported from the North Slope based only on teeth.[10]

Results of the 2005–2007 Period of Fieldwork at Kikak-Tegoseak

The multidisciplinary team led by Paul McCarthy and Tony Fiorillo added a significant body of data that "fleshed out" the paleontological, paleoecological, taphonomic, and sedimentological characteristics of the Kikak-Tegoseak bone bed and its geologic context during the period from 2005 to 2007. The partial skulls and occipital condyles studied thus far support an assignment to the centrosaurine taxon *Pachyrhinosaurus* (see figures 5.7, 5.8). This find makes the Kikak-Tegoseak site home to the farthest northern occurrence of this taxon in a bone bed accumulation. The closest *Pachyrhinosaurus* bone bed located outside of the North Slope is found near Grande Prairie in west-central Alberta over 2,000 miles (3200 kilometers) to the south. The degree of weathering and desiccation of the bones found at the Kikak-Tegoseak site, along with evidence of dermestid beetle activity, indicates that the carcasses were exposed to some subaerial conditions prior to burial. This stands in contrast to the characteristic taphonomic signals found at the hadrosaur-dominated Liscomb and related bone beds.[11]

The addition of a palynologist, Pierre Zippi, to the team by 2007 allowed for a detailed analysis of the sediments and the pollen assemblage contained therein. The result was the formulation of a paleoenvironmental model that placed the pachyrhinosaurs in an "anastomosing splay complex."[12] Broad floodplains were being crossed back and forth by broadly meandering sediment-chocked rivers that spilled over their banks and levees during spring thaw and intense summer thunderstorms. The floodplains between anastomosing channels were well drained and contained occasional and ephemeral bodies of water with no marine influxes. The immediate area was dominated by nonarboreal plants, and the presence of charcoal debris reflects occasional forest fires in surrounding coastal forests. Shallow, incipient soils developed during periodic saturation followed by periods of desiccation.

The soils that developed at and near the Kikak-Tegoseak were formed under a paleoclimate that can be characterized as cool temperate with mean annual temperatures that were greater than 40° F (5° C). Pollen analysis and plant megafossils indicate a cool to warm temperate paleoclimate, and mean annual temperatures of 43° F to 52° F (6 to 10° C).[13] The Arctic today at comparable latitudes experiences cold to frigid polar climates with mean annual temperatures hovering around 18° F (−12° C).[14]

To date, the partial skeletons of at least ten individual *Pachyrhinosaurus* have been identified. The skeletons are highly disarticulated, and some partial skulls were found to be nested against one another. The bones exhibit little or no abrasion but have surface fractures and signs of surface delamination. Concentrations of small, semicircular pits suggest dermestid beetle activity.

The evidence accrued to date points to the presence of a large population of pachyrhinosaurs at a high (> 70° N) paleolatitude with a cool temperate climate, a population that was supported by abundant vegetation, comprised of trees, shrubs and herbs An aggregate of subadult *Pachyrhinosaurus* were drowned in a river by an intense seasonal pulse of water that mixed the carcasses with skeletal debris from theropods and hadrosaurs. The *Pachyrhinosaurus* remains underwent some subaerial drying and beetle scavenging prior to burial.

Close Encounters: Bears, Arctic Rivers, and the Kikak-Tegoseak Site

In eighteen years of fieldwork that covered a good part of Alaska, including the bear-rich Alaska Peninsula, I was fortunate to have only one close encounter with a bear. This strange encounter occurred at the Kikak-Tegoseak site and still elicits a chuckle whenever I recall the incident.

In 2001 Tony Fiorillo and I were assessing a cluster of skulls that a graduate student, James Sammons, had unearthed near the top of the bluff. Tony and I were in the process of helping the graduate student get his bone bed mapping and excavation program started when he discovered the skulls. While Tony Fiorillo and I took some measurements and cleared away debris and brush, the student rejoined two volunteers who were continuing work on another part of the bone bed. At this site, we always stashed our packs and food near the top of the bluff just below the tundra "rug" that stretched for hundreds of miles to the north, south, and west. On every previous visit to this site, we had come across the torn-up ground that marked the food search of hungry brown bears that worked their way north from the Brooks Range each summer, feeding on a favorite snack—the Arctic ground squirrel. This day, a bruin came on our packs, which were sitting between the two teams some 60 feet (20 meters) away. I was the first to see the bear. I immediately called to the other team members to crouch down and stay still. I then called to my graduate student to ask if he had his rifle and pepper spray close at hand. "No," he said. "they are with the packs." "Well," I answered, "I am so glad that the bear is safe from us." Any field scientist that has worked in Alaska for any length of time has his or her memorable bear story.

James Sammons and his team of dinosaur diggers managed to cap off the 2001 field season at Kikak-Tegoseak with another exciting and potentially disastrous episode. James was trying to get as much fieldwork completed as possible and pushed his work into early September, always a dicey thing to attempt on the North Slope. In addition, the boat used to ferry the crew across the Colville to the study site was, by this time, minus the more powerful primary outboard engine due to a mechanical problem. This left the crew with an underpowered engine and no backup. Under the best of conditions, this situation would not have been life threatening; however, weather in the Arctic must always be considered to be in a state of change. James and his volunteers had seen the Colville steadily rising for some three days. Each day it became more difficult to fight the increasing current and make the round-trip from the base camp, on a river bar, to the dig site on the opposite shore and some 115 feet (35 meters) up a steep bluff. In addition, the area of the base camp was becoming smaller as the waters rose and ate up the lower surroundings.

By the fourth day, James realized that this was becoming so dangerous a situation that it was time to evacuate the camp. He used his satellite radio to call pilot Walt Audi of Alaska Flyers. Walt was in the middle of trying to pull out several other customers from the opposite end of the North Slope but decided that James and his crew should get first priority. When Walt arrived, he found that there was no longer an adequate landing zone anywhere close to the base camp. So he flew on to the Deadhorse airport near Prudhoe Bay and notified the Barrow volunteer search and rescue team. The following day, snow flurries and colder temperatures made the Kikak-Tegoseak crew's predicament even more threatening. However, the Barrow rescue team had their hands full, as the flooding rains and sleet had expanded and caught up with others on the North Slope. By this time, James had exhausted the batteries on his satellite radio, so he couldn't alert the Barrow rescue team to the worsening situation. Fortunately, the Barrow rescue team had contacted a CH-47D Chinook helicopter out on a training mission and asked them to assist the Kikak diggers. The U.S. Army Chinook could not transport unauthorized civilians but could carry equipment and fossils. When Major Young and her crew arrived at the Kikak base camp, she was struck by the near desperate conditions and immediately flew to Deadhorse, where she strongly suggested (I am being diplomatic) that the Kikak crew be placed at the top of the rescue team's priority list.

The Barrow Search and Rescue helicopter arrived within an hour to the great relief of the dinosaur hunters. The rising waters had marooned the diggers on a very small part of the river bar. The rescue team had failed to appreciate the extent of the flooding at this point on the river and how dangerous it would have been for James and his volunteers to traverse the raging Colville in an underpowered boat to get to the higher ground across the river. The temperature of Colville River hovers around 40° F during the summer, and immersion, even with a lifejacket, can cause hypothermia and death within less than half an hour. Hypothermia and drowning are the two greatest causes of death among field scientists in the Arctic. Fortunately, all ended well with James and his three volunteers, who arrived safely at Deadhorse with all of their gear and fossils, thanks to Major Young and the efforts of the Barrow rescue team.

The focus of the discussion of the paleo-Arctic dinosaur record of Alaska now shifts from the North Slope to south-central Alaska, nearly 200 miles (324 kilometers) northwest of Alaska's largest city, Anchorage.

Tracking Dinosaurs in Denali National Park, 2001–2008

Just north of the northern end of the Talkeetna Mountains near where the hadrosaur and hadrosaur skull had been discovered in 1994 is one of the most spectacular national parks and preserves to be found in Alaska. Denali National Park has now become the venue of the most recent dinosaur discoveries in Alaska.

If you asked visitors to Denali National Park and Preserve what kind of large animals could be tracked within the park, they would most certainly say bears and/or caribou. However, they might also now add dinosaurs owing to a program of paleontological resource evaluation that Tony Fiorillo and I began in cooperation with the National Park Service in Alaska in 1999. Tony and I first worked with Russell Kucinski and Vincent Santucci on parks on the Alaskan Peninsula and in the Yukon-Charlie River National Preserve and then turned our attention to Denali National Park and Preserve. Earlier geologic mapping and sedimentological studies had established that the lower member of the Cantwell Formation was of the right age (Late Campanian to Early Maastrichtian) and had the correct alluvial environments for dinosaur hunting.[15] In addition, the Cantwell sediments were deposited at a high paleolatitude, somewhere close to or above (63–70°N) the paleo-Arctic Circle.[16] Tony and I convinced the Denali Park geologist, Phil Brease, that there was a high probability that dinosaurs could be found in the Late Cretaceous lower Cantwell Formation. In addition, the Denali Park Road went right through well-exposed outcrops of the Cantwell near Igloo Canyon and the Polychrome Pass area. This gave relatively easy access for prospecting trips in 2000, 2001, and 2002. Tony Fiorillo, Phil Brease, and I made a concerted effort to prospect several well-exposed outcrops of the lower Cantwell Formation north and northeast of the Denali Park Road just to the northwest of Igloo Canyon and near Polychrome Pass. These efforts met with no success, but Tony and I were nevertheless encouraged by what we saw, and we concluded that it was just a matter to time before the first dinosaur remains would be found. And that proved to be the case: in 2005, a geology student participating in McCarthy's field geology course in Denali Park discovered the first dinosaur footprints in Igloo Canyon. Multiple bird tracks were subsequently discovered at another locality. These initial discoveries were followed by a series of finds, including a natural cast of a manus track attributed to a pterosaur. This evidence of a flying reptile (ichnogenus, Pteraichnus) was collected in 2008 and would represent the highest latitude occurrence of this group in North America.[17] As of 2006, twenty trace fossil sites were located in the Igloo Canyon and Double Mountain areas.[18] The Cantwell Formation is thousands of feet thick and crops out over a large part of the central Alaska Range in and around Denali National Park and Preserve. Fossil vertebrate tracks have been located in Denali State Park just to the east of the national park and preserve by Kevin May.[19] I am sure that there will be more dinosaur, bird, and other vertebrate tracks found in the Cantwell Formation. It is just a matter of time and effort before more are located in and out of the Denali National Park and Preserve. It is also highly likely that body fossils will be found in the Cantwell Formation. The lower part of the upper volcanic member may very well contain body fossils trapped in volcanic ash, and what is very exciting is that the Cretaceous-Paleogene boundary may ultimately be found in this member.

Having considered the extensive body of evidence that has been unearthed over some twenty-years years showing that there were dinosaurs in paleo-Arctic Alaska, I now want to turn to the central question of this chapter. Were paleo-Arctic dinosaurs residents or long-distance seasonal visitors?

Arctic Dinosaurs: Content Homebodies or Solar Vagabonds?

As the abundant and widespread record of dinosaurs was being uncovered in the Arctic, several paleontologists adopted the position that these dinosaurs, lacking food and/or the physiology or adaptations to cold such as warmth-retaining pelage to withstand the cold, would not have been capable of survival during the colder winter months. Therefore, they concluded that paleo-Arctic dinosaurs must have been long-distance or primarily latitudinal migrants.[20] They, like their avian cousins, would have migrated en masse over thousands of miles of latitude to the food-rich paleo-Arctic for a summer's feast and then would have returned to their southern home grounds following a well-defined route as winter approached. This seasonal migration would have primarily taken the form of a latitudinal traverse, but it would not necessarily have been a straight-line journey. Part of this model or preliminary hypothesis was based on the tentative conclusion that dinosaurs either were not endotherms or were inefficient endotherms and therefore could not have tolerated the lower light values and winter temperatures or the decreased food availability polar winters would have presented.[21] Proponents of the long-distance model for North America envisioned seasonal migrations over distances of at least 1,800 miles (3,000 kilometers) round-trip. However, if various paleo-Arctic plate-tectonic solutions to seafloor expansion are taken into account, the distances they would had to travel would have amounted to 2,300 to 5,600 miles (6,000 to 9,000 kilometers).[22] When pressed to cite living, terrestrial, nonavian analogs that the dinosaurs might be compared to, the long-distance migration proponents cited Arctic barren-ground caribou in Alaska and Canada. Several modern terrestrial mammals such as wildebeest, Burchell's zebra, and several types of gazelle are known for their long-distance movements. However, barren-ground caribou are the only modern terrestrial mammal that fit the size, reside in the Arctic, and exhibit behavior that can properly be defined as "migratory."[23]

Gregory Paul, a consistent opponent of the long-distant seasonal migration model, countered with a penetrating analysis. Paul's counterargument was primarily based on physiological principles and a comparative energy budget for caribou that was cited as a hypothetical hadrosaur model for their migrations. Paul concluded that the energy costs would be too great for the less efficient hadrosaur adult to migrate such great latitudinal distances just to reach more productive southern grounds. R. McNeill Alexander has subsequently added more data and analysis, based on physiological grounds (energy demands and postural efficiency), that emphasizes that the distance and the dangers posed en route to migrating animals would have been too great.[24] He concludes that his data and analysis points to long-distance vertebrate migrations of 6,000 miles (10,000 kilometers) or more, which would not have been beneficial to walkers. Annual round-trip migrations over these distances would only have been worthwhile for fliers and endothermic swimmers. By extension, long-distance latitudinal migration by hadrosaurs or ceratopsians is highly questionable. We can never know for sure whether the basic assumptions of Paul and Alexander regarding biomechanics and physiology are correct, since we will never be able to make direct measurements on living dinosaurs. The use of laboratory models made from fossil bones and the study of living surrogate animals, when done within logical limits, is a reasonably valid approach in paleontology. After all, it is our only choice. However, paleontologists can use geographic distribution, density of fossils, and taxonomic diversity to check their conclusions when they have a reasonably good fossil record.

It has been pointed out that the composition and geographic distribution of Late Cretaceous dinosaur faunas strongly suggest that endemism and provinciality characterized their evolution after an earlier geographic radiation.[25] A recent study of stable isotopes in hadrosaur teeth represents an independent line of evidence that is relevant to this working hypothesis.[26] Stable isotope ratios of oxygen and carbon accumulate in tooth enamel as a result of eating and drinking in a particular paleoenvironment. When analyses were made for a series of samples collected from different geographic locales along the Western Interior Seaway, little to no overlap of the stable isotope ratios was noted after diagenetic (rock-forming) influences were eliminated. If the hadrosaurs that were sampled had been making regular long-distance migrations from north to south, the isotopic signatures would have been obscured (overlapped) rather than distinct. Even though neither of these analyses can be considered unequivocal, they do offer two more lines of evidence in opposition to the long-distance migration hypothesis.

North Slope and Other Arctic Dinosaur Records Weigh In on the Migration Debate

By 2007, analysis of four remarkably rich bone beds on the North Slope had been completed by me in concert with colleagues Kevin May, Tony Fiorillo, and Paul McCarthy. We established the presence of small-bodied taxa (a pachycephalosaur and at least four theropods), concluded that concentrations of individual animals found in the bone beds suggested that hundreds if not thousands of individuals died en masse, determined that the bone beds were greatly dominated by a single species of herbivorous dinosaur (either a hadrosaur or a ceratopsian) and that the Liscomb and several closely related bone beds were dominated by juveniles with minor numbers of adults and subadults. These conclusions inform the debate on the possible migratory behavior of paleo-Arctic dinosaurs.[27] That the beds were dominated by juveniles of a single herbivorous species strongly suggests that the individuals contained in the bone beds probably engaged in some form of pre-mortem aggregation and interaction. Although the data could be interpreted as evidence for migration, it does not necessarily support a long-distance migration hypothesis. The presence of so many young would greatly restrict the distances that the adults could cover because the juveniles would have difficulty traveling great distances due to their body size and immature physiology. Taken as a whole, this record refutes rather than supports the long-distance, latitudinal migration theory.

A very exciting recent discovery of a diverse Late Cretaceous (Late Maastrichtian) dinosaur fauna at Kakanaut in the Koryak Upland (see figure 2.1) in the far east of Russia has added the presence of eggshell from hadrosaurs and nonavian theropods to the record of the paleo-Arctic.[28] In addition, an even more recent find of abundant hatchling- to nestling-sized hadrosaurid fossils at a locality in the Grande Prairie area of west-central Alberta attests to the presence of a dinosaur nesting site.[29] This site is of Late Cretaceous age (Campanian) and was at or just below the paleo-Arctic Circle. In addition, deposits on the North Slope of Alaska have yielded a partial neonate or hatchling ornithopod humerus.[30] Dinosaur bones from Bylot Island in the Canadian Arctic (see figure 2.1) are attributable to juvenile hadrosaurs.[31] When all of this evidence is taken into account, it strongly questions the hypothesis proposed by some researchers that paleo-Arctic dinosaur migrated yearly over the long distances. Even if adult hadrosaurs and ceratopsians could have traversed such distances in a single year, juveniles would most probably have been physiologically and anatomically unable to do so.[32]

It should also be pointed out that the presence of small dinosaurs such as troodontids, dromaeosaurs, pachycephalosaurs, and ankylosaurs in the Late Cretaceous of the Arctic and high subarctic latitudes also calls into question the long-distance migration hypothesis.[33] The behavior, anatomy, and energetics of these types of dinosaurs were quite different from that of the large-bodied adult hadrosaurs and ceratopsians.[34]

Are Caribou a Good Modern Analog?

As noted, some proponents of the long-distance latitudinal migration hypothesis cite caribou as a modern analog to explain paleo-Arctic dinosaur distribution and survival.[35] Several extensive studies of caribou migration patterns in the Arctic and subarctic of Alaska and Canada do not support using this mammal as an analog for long-distance latitudinal migratory behavior even though it is the longest ranging modern terrestrial animal.[36] Caribou do not traverse great latitudinal distances in great numbers over the year. Rather, they move in complex back-and-forth patterns that cover almost as much longitude as latitude. They often move from interior to coastal areas and come together in great numbers for mating and calving, but they also break into smaller groups, including harem and nursery groups, to forage.[37]

The cumulative data strongly support the argument that the Late Cretaceous dinosaurs of the paleo-Arctic were primarily homebodies (residents). Seasonal migrations most likely occurred, but they would have principally taken the form of longitudinal (east-west) movements from interior and upland areas to coastal habitats and back. Latitudinal movements were most probably short (hundreds of miles or kilometers) north-south displacements of southern populations.[38] The interior and upland areas would have provided a profusion of food sources that renewed each year during the summer. The coastal areas would have provided warmer winter temperatures and, in the case of shallow bays and beaches, food sources not available in the interior. Foods such as seaweeds and possibly sea grasses would have been within reach, just as seaweeds provide fodder for caribou and domestic sheep along Arctic coastlines today.[39] Offshore islands could have provided protection for the young just as they do now for caribou.[40] However, not all of the populations of paleo-Arctic dinosaurs would have had to migrate to find sufficient food sources. That supportive food sources were in short supply during the paleo-Arctic winter is a key claim of the advocates of long-distance migrations. But large numbers of barren-ground caribou overwinter in the interior of Canada and Alaska under much harsher weather conditions than mid- to Late Cretaceous paleo-Arctic dinosaurs would have experienced.[41] These caribou feed on a variety of living green plants that transit the harsh Arctic winters. These plants foods include "the meristem of grasses growing under the snow blanket, such as *Festuca* or *Deschampsia;* in the green leaves of *Empetrum;* in the lower blades of sedges; and in the thin stems of green, creeping *Equisetum scripoides.*"[42] All of these plant groups existed in the Late Cretaceous. However of these, only the Sphenophyte *Equisetites,* a close relative of *Equisetum,* the horsetail, has been identified in sedimentary rocks that are closely associated with the dinosaur bone beds. The likelihood that all of these plant groups would have been even more abundant and joined by others during the paleo-Arctic winters is strongly suggested by the much milder, temperate to cool temperate, climates that prevailed during the Cretaceous compared to the Arctic today.

If paleo-Arctic dinosaurs were not Arctic residents and did engage in long-distance, seasonal, latitudinal migrations, then they most likely don't have a modern terrestrial analog. If paleo-Arctic dinosaurs were year-round residents, what overwintering strategies might they have engaged in?

Hypothetical Overwintering Techniques

Denning and hibernation are overwintering strategies that could have been available to Late Cretaceous Arctic dinosaurs. Although no direct evidence for this type of behavior has been discovered in dinosaurs from Alaska or the rest of the Arctic, evidence for denning by mid-Cretaceous dinosaurs has been reported from Montana.[43] Microscopic osteological patterns (lines of arrested growth) observed in fossil bones from the Early Cretaceous of Victoria, Australia, suggest that some of the dinosaurs, such as the theropod *Timimus,* hibernated periodically. Thus far, preliminary studies of Alaska's dinosaur bones have not detected lines of arrested growth. A contemporary of *Timimus,* the basal ornithopod *Leaellynasaura* was also found to be devoid of them. However, *Leaellynasaura* possessed large eyes (orbits) and enlarged optic lobes that could have allowed it to adapt to life at high latitudes during the low-light conditions of the winter.[44] The paleo-Arctic fauna of Alaska and Chukotka also include basal ornithopods and the unusually large-eyed theropod (*Troodon formosus*). In addition, the North Slope of Alaska has yielded *Alaskacephale,* a pachycephalosaur—another taxon with relatively large orbits that could certainly have likewise adapted to low-light during the paleo-Arctic winter.[45]

What Tracks and Trackways Can Tell Us about Migratory Behavior

If Cretaceous dinosaurs were migrating long or short distances, they should have left a record of their pathways as concentrations of tracks. Just what can be learned from the study of dinosaur tracks and trackways? Individual tracks or footprints are fossils and evidence that a dinosaur was present at some time in the past, just as a bone or tooth attests to such a presence when found. In most cases, a fossil footprint, also known as an ichnite, represents evidence that has not moved over time while a bone or tooth could have been transported to the site under study and therefore be out of context. Dinosaur tracks and trackways are now proving to be more numerous than bones and teeth.[46] If a footprint is part of a series or trackway, then it is easy to determine that it is in place and represents a very small segment of time—the amount of time it took the dinosaur to walk over the surface before it turned into rock.

A footprint or ichnite can reflect the basic shape of the dinosaur's foot, the number of toes, and to some extent, the mechanics of how the dinosaur carried its weight. However, so much more can be learned from a trackway or a series of trackways. A group of trackways can reflect the range of ages of the population that formed them and whether the dinosaurs were traveling in the same direction and therefore exhibiting gregarious behavior of some sort, such as herding. A concentration of tracks and trackways combined with evidence of trampled and dried-out skeletons that is associated with accumulations of salts and mudcracked sediments could be attributed to a drying waterhole during a drought. Trackways can give estimates of stride length while the animal walked or ran, and indicate whether the dinosaur carried its tail above the ground or drug it while engaged in locomotion. In the present state of ignorance, dinosaur "trackers" cannot assign particular tracks or trackways to genera or species but can recognize consistent patterns that reflect major groups, or clades, such as ornithopods, ankylosaurs, ceratopsians, sauropods, and theropods. Because of this limitation of taxonomic assignment, dinosaur ichnology has not always been given proper emphasis or value in paleontological studies. As more trackways are discovered and studied using a set of standardized protocols, trackways promise to have more impact on the study of dinosaur paleoecology, paleogeography, and paleoethology (behavior). Several aspects of individual behavior can be discerned from tracks and trackways. Resting, coping with an injured limb, or missing a toe, can be detected by careful study of tracks and trackways. One highly intriguing aspect of dinosaur behavior that can be elucidated with ichnological studies is the linked behavior of predators and prey. Herbivorous and carnivorous dinosaur trackways have been found on the same surfaces where one or the other shows a rapid change in direction that could be interpreted as predator-prey behavior.[47]

The evidence of dinosaurs that has accrued over the last forty-seven years in the circumarctic region very clearly establishes that not only were Cretaceous dinosaurs widespread in the paleo-Arctic but that they thrived there. Now it is time to discuss what the western Arctic region was like in the mid- to Late Cretaceous and how dinosaurs may have adapted to this part of the world.

6 THE ARCTIC DURING THE CRETACEOUS

North America and the western Arctic as experienced by dinosaurs and their kin were much different places than today. Today we find the planet dominated by large continents with interiors that are relatively far removed from the ocean margins. The oceans are now at what is called a low stand, because significant amounts of water are locked up as ice caps at both poles. Oceans act as great moderators of temperature and moisture. They are conveyors of heat energy, transferring energy to the colder land masses during winters and acting as coolers when the continents heat up during the summer.[1] The oceans act as reservoirs of moisture by recycling water to the continental surfaces as precipitation. The pattern of precipitation is directly linked to the flow of heat energy and to the patterns of high and low pressure. Temperature and precipitation directly control the distribution and types of vegetation that are available to terrestrial animals such as dinosaurs.

Alaska and the Western Interior Seaway

To understand the ancient world that dinosaurs of Alaska and the rest of the western Arctic dealt with, one must go back to the Cretaceous of one hundred to sixty-eight million years ago and look at the distribution of land and sea. Today, it is hard to envision the large continent of North America as composed of two to three subcontinents—a smaller and skinny western landmass, and at various times during the Cretaceous, one or two broader landmasses to the east. These landmasses were separated by shallow interior seas. Alaska during most of the Cretaceous was at the northern end of an island-like subcontinent that was thousands of miles long and hundreds of miles wide along most of its length.[2] It was bordered on the west by what is now the Pacific Ocean. This western ocean reached depths over 10,000 feet (3,000 meters). The eastern side of this long narrow island was washed by a shallow interior sea that was less than a 1,000 feet (300 meters) deep at its deepest points. This North American interior seaway has been given a variety of names in the geologic literature that reflect the rich marine record of fossil life that it nurtured, but I refer to it as the Western Interior Seaway. It stretched from the paleo-Arctic southward to what is now the Gulf of Mexico and reached 1,000 miles (1,600 kilometers) at its widest part.[3]

Computer models indicate that seaway was often subject to vertical mixing over the year, and this helps to explain the rich marine faunas and floras that fossil collectors have detailed for over a hundred years. The southern end of this seaway was part of an even more extensive warm, life-rich, mega-seaway, called the Tethyan Seaway, that stretched almost halfway around the waist of the earth.[4] The proximity of the ocean and its warm currents reduced the harshness of the paleo-Arctic winters and provided abundant precipitation throughout the year—a set of conditions supportive of overwintering for at least some dinosaurs. Several computer models for the Western Interior Seaway have indicated that it was affected by large gyres that brought warm southern waters northward.[5] The presence of a connection to tropical warm waters at the its extreme southern end and the natural moderating effect of water masses on three sides gave Cretaceous Arctic Alaska a climate similar to that found today some 2000–3,000 miles (3,200–4,800 kilometers) to the south along the coasts of northern California, Oregon, and Washington. Mean annual temperatures ranged from 7.0 to 13° C (42 to 52° F) in the mid-Cretaceous to 2.0 to 8.0° C (33 to 43° F) at the end of the Late Cretaceous. This compares with mean annual temperatures of –12 to –8.0° C (10 to 18° F) at similar latitudes in the Arctic today.[6] Unlike now, the Arctic Ocean of the mid- to Late Cretaceous was largely ice free.[7] It is important to note that in the Late Cretaceous, the North Slope dinosaur localities of Alaska were at least 70–75° N and may have been near 85° N, some 15° farther north than today.[8]

Shorelines, Deltas, Dinosaurs, and Coal

The western edge of this great seaway provided long stretches of shoreline and beaches with warm, sun-bathed offshore shallows that in places extended for miles offshore. Great upheavals of the island's interior formed the ancestral Rocky Mountains and Brooks Range. These rapidly building mountains, driven by the forces that spread the ocean floors and send continental plates colliding with one another, resulted in a landscape dominated by large, meandering river systems that carried huge loads of sediment eastward.[9] These sediment- and nutrient-rich rivers formed massive deltas as they spilled into the shallow margins of the seaway. The deltas and their labyrinthine systems of rivers and sloughs fostered dense concentrations of aquatic and semiaquatic plants that, in turn, supported great concentrations of herbivorous dinosaurs. Hadrosaurs dominated these delta faunas. Ceratopsians were found in smaller numbers but were found in greater numbers on drier uplands farther inland. Landward, floodplains and delta margins supported dense evergreen and deciduous forests with an understory of ferns, cycads, and some flowering herbaceous plants that provided abundant fodder for a variety of ceratopsians.[10] Coastal swamps formed behind baymouth sandbars while mires formed farther inland, supporting dense masses of aquatic plants.[11] These wetlands became traps for local and transported plant remains that, when buried deeply enough, would eventually be turned into a variety of coals.[12] The Cretaceous is a time when most of the great western and northwestern North American coal deposits formed.

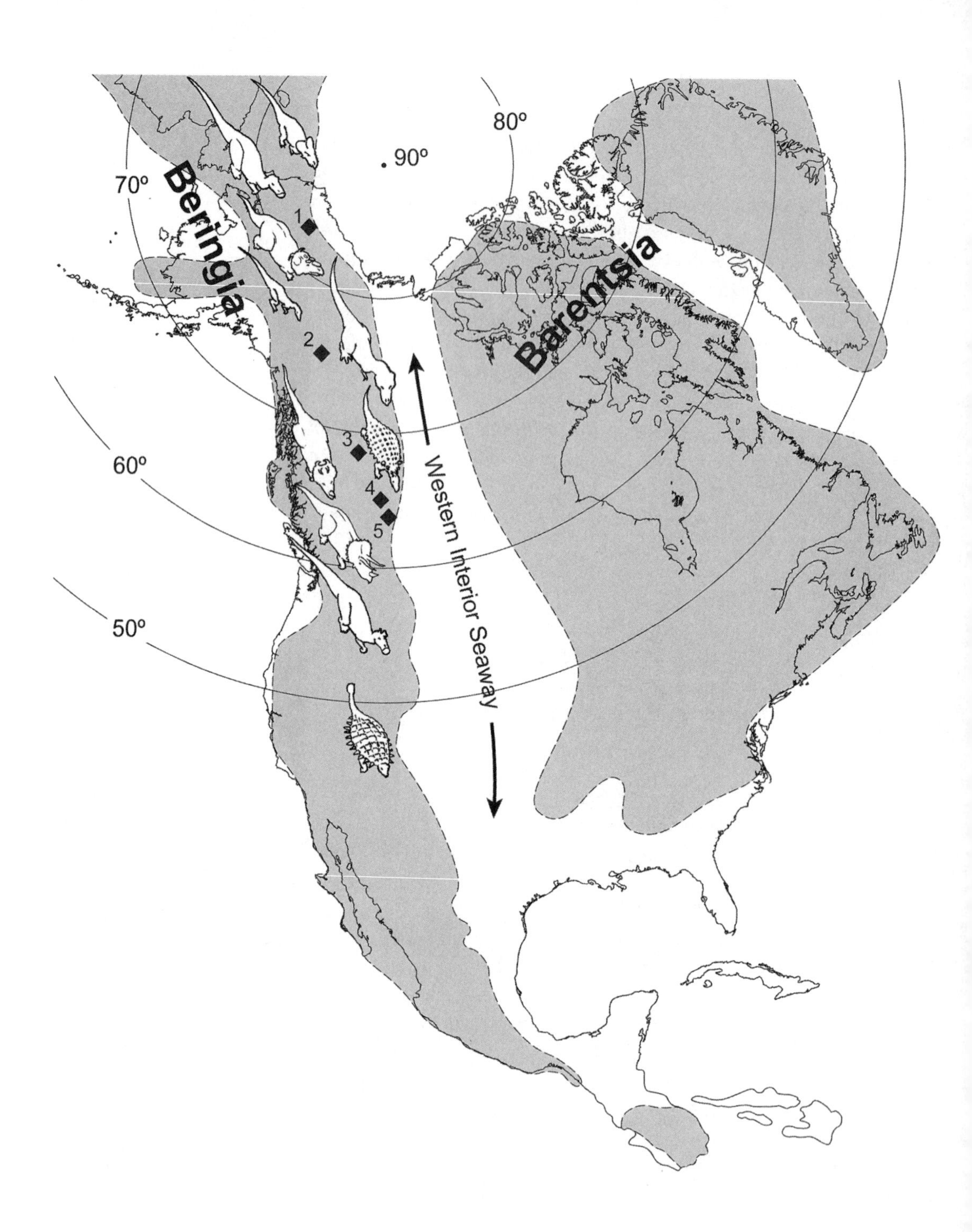

The Importance of Cretaceous Coastlines as Dinosaur Habitat

Many researchers have noted that hadrosaurs were supremely adapted to exploit the rich vegetation to be found along great rivers, on deltas, and lake shorelines during the Cretaceous. However, only a few seem to have appreciated the potential importance of ocean shorelines as feeding grounds and especially as winter oases in the paleo-Arctic of North America.[13]

The Cretaceous Western Interior Seaway afforded long stretches of complex shorelines that would nurture and harbor kelp and other seaweeds that floated in the near-shore shallows and would wash up and accumulate on the sandy beaches.[14] Hadrosaurs and ceratopsians, with their tooth batteries, would have been well adapted to feeding in this abrasive-rich environment, just as modern caribou and sheep feed on a variety of seaweeds that accumulate along beaches in the Arctic. However, unlike quadrupedal caribou and sheep, bipedal and semibipedal hadrosaurs could have waded offshore to feed on seaweeds and stands of sea grasses in the subtidal zone, especially in the protected interdistributary bays of the great deltas (see figure 6.2). Ceratopsians with their obligate quadrupedal locomotion would have been more limited in these offshore habitats. Sea grasses appeared during the Early Cretaceous (Albian), and by the Late Cretaceous may have invaded the Arctic.[15] These ocean shoreline resources would have been particularly important during the colder and sun-starved winter season. The extensive and complex shorelines of the Western Interior Seaway and its circulation patterns would have offered warmer temperatures, more abundant food sources, and even special shelter for young hadrosaurs during the winter and nesting refuges during the summer.[16]

The ceratopsids probably were not as numerous along ocean shorelines as the hadrosaurs. Ceratopsids could have reaped some of the environmental rewards that lured the hadrosaurs, but their range of edible vegetation would probably have been much more limited. Although *Pachyrhinosaurus* and kin could have been seen cutting stringers of kelp into bite-sized pieces on occasion, these ceratopsians would probably not have congregated to the extent they would have when they were farther inland where their primary food sources such as cycads, ferns, and young deciduous conifers would have been more common.

Fig. 6.1. Cartoon map of Beringia, Barentsia, and the Western Interior Seaway during the Late Cretaceous with emphasis on the possible long-term migration route, Western Cordillera region. Important fossil localities are designated by numbers: 1) Liscomb Bone Bed, North Slope, Alaska 2) Whiskers Lake, Yukon Territory, Canada 3) Peace River, British Columbia, Canada 4) Grande Cache-Smoky River Coal Mine, Alberta, Canada 5) Drumheller, Canada. Paleogeographic map based on that of Smith, Hurtley, and Briden (1981). *Credit: David Smith.*

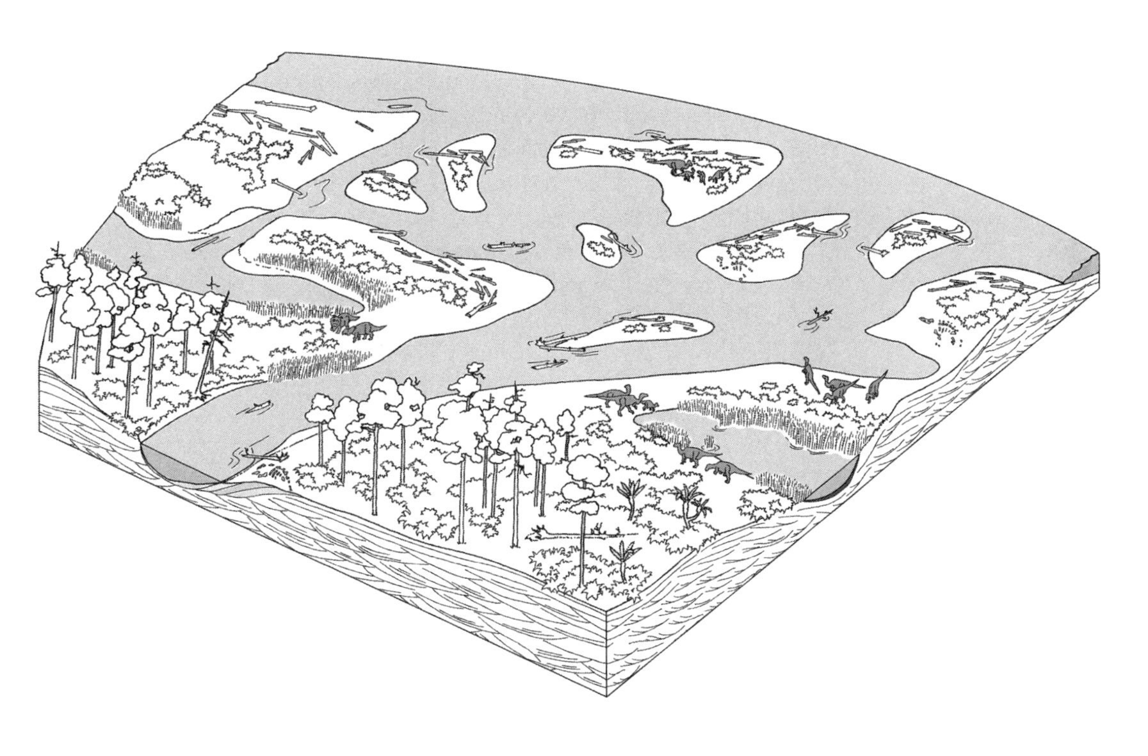

Fig. 6.2. Cartoon depicting variety of shoreline habitats during the Late Cretaceous in North America along the western coast of the Western Interior Seaway. *Credit: David Smith.*

The Western Interior Seaway was a life-rich epicontinental seaway fed by a system of massive rivers and an ocean circulation that brought waterborn nutrients into the water column—the nutrients sustaining all manner of organisms.[17] The shorelines and near-shore areas of the seaway most likely held dense "gardens" of kelp, green algae, and perhaps sea grasses. Thus, this array of plant-rich environments would have provided lush feeding grounds for herbivorous dinosaurs that, in turn, would have provided food for a diverse assortment of predators. These shorelines with their complex habitats, food sources, and winter-moderating attributes would have been very enticing destinations and pathways for seasonal migrations.

This account of the possible exploitation of high-latitude shoreline habitats by dinosaurs is speculative. Good science requires that the geologic record be consulted to determine just how speculative it might be.

What Does the Geologic Record Say?

The important role played by the complex shorelines of the Western Interior Seaway in the lives of Cretaceous dinosaurs becomes apparent when the record of Cretaceous dinosaurs of North America is scrutinized. In 1979, Jack Horner of Montana State University compiled an impressive list of dinosaur remains that were found in marine sediments from many different parts of the seaway.[18] Close anatomical inspection of the remains led to the conclusion that these were not marine or truly aquatically adapted animals. They were anatomically similar to the fossil remains found in terrestrial rocks of the same age. Subsequent studies of the distribution of hadrosaur, ankylosaur, and ceratopsian remains have corroborated the close association of many members of all three dinosaur lineages with wet coastal and near-coastal environments.[19] However, more recent analyses point out that only nodosaurid ankylosaurs and representatives of the centrosaurs join the hadrosaurs in coastal and near-coastal habitats.[20]

Could dinosaurs end up in deltaic and shallow water marine sediments if they were not living in or near these habitats? One reasonable explanation for finding the remains of dinosaurs in coastal and even shallow marine environments can be constructed using contextual (taphonomic) data. These individuals died far upriver and were transported, postmortem, to the sea. Dinosaurs that died upstream may have drowned under a range of circumstances, including while attempting to cross rivers in flood stage. After drowning, their bodies would have been made buoyant by gases generated internally by early bacterial decay. The inflated carcasses then would have floated downriver to the coasts or into the sea, thus earning the appellation of "bloat and float" dinosaurs.

A very fine example of the bloat and float process that resulted in the burial of a dinosaur carcass in a marine setting was discovered in the Talkeetna Mountains of south-central Alaska. The site is some 90 miles northeast of Anchorage (see figures 1.2 and 2.1).[21] The partial postcranial skeleton of a hadrosaur juvenile was found in pyrite-rich mudstone (figure 6.4) assigned to the Late Cretaceous Matanuska Formation. The surrounding mudstone also contained a diverse assemblage of marine invertebrates. The invertebrate fauna included abundant ammonites and an array of

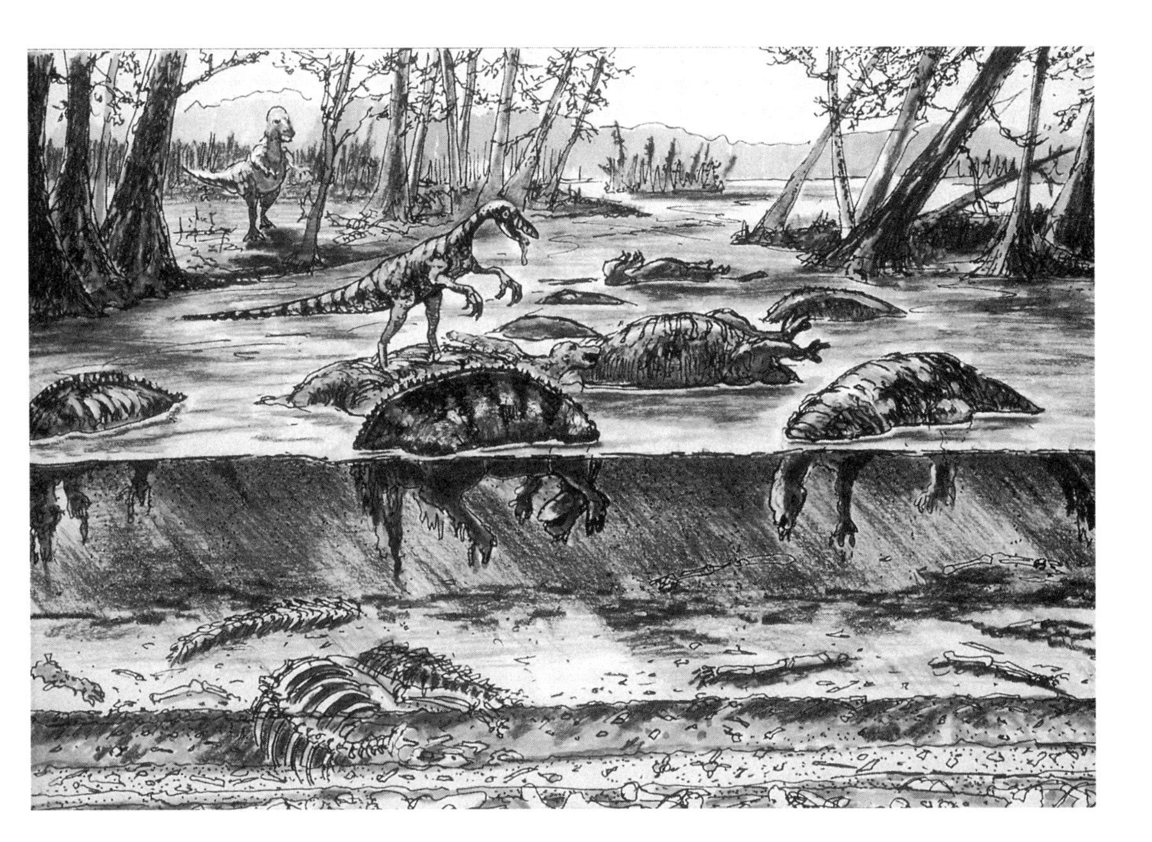

other mollusks as well as a solitary coral. Numerous teeth of sharks and teleost fish were recovered from closely associated sediments. Various marine microfossils such as radiolaria and foraminifera combined with diagnostic ammonite taxa such as *Muramotoceras* point to an outer shelf marine environment of Late Cretaceous (Middle Turonian) age for the enclosing rocks. Pollen and wood fragments are attributed to terrestrial taxa.

Fig. 6.3. Cartoon depicting "bloat and float" with carcasses of juvenile *Edmontosaurus* in backwater area. *Troodon* in foreground with a tyrannosaurid in background. Based on content of Liscomb Bone Bed. *Credit: Tom Stewart and University of Alaska Museum of the North, Fairbanks.*

The skeleton is partially represented by seventy skeletal elements from all four limbs, one forefoot, two hind feet, several vertebral regions, ribs, the shoulder girdle, and possibly the pelvis. The caudal vertebrae and parts of the appendicular skeleton are articulated. The rest of the recovered skeletal elements are found within a yard or so. Some of the bones exhibit depressed fractures and scars that are interpreted as evidence of scavenging, most likely by a marine reptile such as a mosasaur. This find is highly significant, not just because its paleoenvironmental context is well documented and because of its implications for the phenomenon of bloat and float but also because it represents one of the earliest records of hadrosaurs in the paleo-Arctic of North America.

Present-day studies along rivers and lakes have revealed a number of variations on the bloat and float phenomenon.[22] Groups of herding animals such as cattle, wildebeests, and caribou have been drowned while attempting to cross rivers or swollen streams. The animals were either migrating or being pursued by predators. Drowning was either caused by the crush of panicky individuals or the hydrodynamics of flooding or both. Drowning has occurred at both broad parts of channels and at narrow and

Fig. 6.4. Concretion from Matanuska Formation, Talkeetna Mountains, Alaska, with partially prepared, pyrite-rich metatarsal and phalanges from the right pes of a juvenile hadrosaur. Scale = centimeters. *Credit: Kevin May.*

steep portions of a river. A very interesting variation was observed along the Yellowstone River in Montana. A herd of forty cows broke through the thin ice that formed near shore during the winter and became trapped under it. Their bloated carcasses were found the next spring distributed along the river's course. These modern studies have helped to shed light on the formation of mass-death bone beds such as the Liscomb and Kikak-Tegoseak bone beds of the North Slope, as well as bone beds in Alberta and Montana.[23] There is little doubt that bloat and float had a role in forming some of the concentrations of dinosaur remains at or near coastal areas. However, the overall importance of this mechanism in the record of bone beds in the Late Cretaceous awaits further research and analysis.

It should be clear that Cretaceous shorelines demand much more research if we are to understand the ancient world of Arctic dinosaurs. Hopefully, you now have a mental picture of the physical environment that western paleo-Arctic dinosaurs operated within. It is now time to explore how the various dinosaurs "made a living" and related to one another.

Western Paleo-Arctic Terrestrial Ecology: The Dinosaur Food Chain

What do the best fossil records from the paleo-Arctic tell us so far? If we could go back in time to the pieces of crust that now bear the names of Alaska and Koryakia in northeastern Siberia, what would we see? Most children and other dinosaur enthusiasts think of the Cretaceous period as the time of *Tyrannosaurus rex* and "raptors" such as *Deinonychus* and *Velociraptor.* Today mention of the Serengeti Plains of Africa conjures up images of stalking lions, leopards and the fleet-footed cheetah. Carnivores loom large in our minds and can erase the sea of grazers and browsers that support them and fill the background. Perhaps evolution has wired our brains to be on the lookout for carnivores. Yet it is two groups of herbivores that by sheer numbers, geographic distribution, and evolutionary diversity rightly claim to be the proper hallmark of the Late Cretaceous: the hadrosaurs and the ceratopsians.

In discussing any food chain, the discussion must start with the primary consumers, or herbivores. Since the focus of this book is on dinosaurs, the role of insects and other primary consumers will not be dealt with. The present Cretaceous record of herbivorous dinosaurs in Alaska is highly skewed towards hadrosaurs. Ceratopsians are present in some abundance, but nothing like their ornithopod cousins the hadrosaurs (see plates 2, 6 and 7). Hadrosaurs are often popularly characterized as the "cows of the Cretaceous period." I would amend this statement: hadrosaurs and ceratopsians were the "bison" of the Cretaceous. That is to say, they most probably formed herds and were both grazers and browsers capable of eating and digesting some of the toughest vegetation, including horsetail rushes, ferns, and cycads, that ever existed on this Earth. It is clear that both of these dinosaur clades (lineages) were superbly equipped herbivores and therefore were at the base of the Cretaceous food chain as primary and opportunistic consumers. The hadrosaurs made their appearance near the beginning of the Late Cretaceous and the horned dinosaurs, or ceratopsians, in the Early to mid-Cretaceous. By the end of the Cretaceous, both groups had exploded in numbers and diversity, accounting, for 65–85 percent of the taxa of large vertebrates that made up ancient terrestrial communities in Asia and North America.[24] When Alaska's record is compared to and correlated with that of Alberta, Canada, the great dominance of hadrosaurine duckbills found in the paleo-Arctic of Alaska strongly suggests that may not be a valid sampling of the original fauna. However, it is also clear that there is a great deal still to be learned about the paleo-Arctic of Alaska. The present record is a preliminary one.

Key Herbivores: The Record of Hadrosaurs and Ceratopsians in Arctic Alaska

In the Late Cretaceous of Arctic Alaska, Hadrosaurinae, the noncrested subfamily, is fairly well represented by cranial and postcranial skeletal elements and presently accounts for the vast majority of specimens. The presence of the other hadrosaur subfamily, the Lambeosaurniae, is represented by a few postcranial bones from several localities, and identifications are problematic. The only member of the Hadrosaurinae recognized thus far is *Edmontosaurus.* No species assignment has been made yet.[25] A species assignment is difficult because juvenile individuals greatly dominate the collections and there is a great paucity of material from full adults, especially of critical components such as skulls. The great bulk of the skeletal material that has been collected (over six thousand skeletal elements) was recovered from bone beds and probably represents an excellent population sample of a single species. Unfortunately, there is no skeletal dataset to be found in other collections that allows for a proper comparison.

Ceratopsians are represented by two fairly complete skulls and parts of at least eight others. These and a variety of postcranial elements have been assigned to the genus *Pachyrhinosaurus* (see plate 7 and figure 5.8 and 6.7).[26] This northern short-frilled genus ranges from Arctic Alaska to southern Alberta, while others of its subfamily, the Centrosaurinae, range from southern Alberta to northern Montana along western North America.[27] In addition to the *Pachyrhinosaurus* skeletal material, a partial horn core and incomplete femur have been tentatively assigned to ceratopsids. An isolated occipital condyle was tentatively assigned to the genus *Anchiceratops*, of the subfamily Chasmosaurinae.[28]

The Late Cretaceous record of herbivorous dinosaurs in the rest of the Arctic that has accrued since the 1980s is similar to that of Alaska except that the fossils are too fragmentary to warrant identification beyond family or subfamily level. It is interesting to note that the record of fossil tracks and trackways (ichnites) in the western paleo-Arctic points to the possibility of abundant ankylosaurs and ceratopsids joining the ubiquitous hadrosaurs.

Jaws and Teeth Tell the Story

Skull features reveal that hadrosaurs and the highly evolved (derived) ceratopsians (Neoceratopsia) exhibit convergent evolution in a system of tooth batteries that occupied both upper and lower jaws.[29] These tooth batteries comprised a series of tightly packed tooth rows, each of which held a sequence of several teeth. The sequence of teeth moved upward, like a conveyor belt as the uppermost tooth in each row was worn down. This assured that teeth would be replaced as they were worn down and the nubbins ejected. Hadrosaurs and ceratopsians represented an acme in the evolution of vertebrate teeth and jaw mechanics.[30] However, the hadrosaurs differed from their horned "dental-mimics" in several important details.

Hadrosaurs had smaller individual teeth with a single root. Ceratopsids had proportionally larger individual teeth with a double root that was split at 90 degrees to the length of the jaw. Hadrosaurs could have had a total of eight to nine hundred teeth in their mouth as an adult, while ceratopsids had far fewer—only 30 to 60 percent as many. Hadrosaurs had three times as many teeth in contact along the chewing plane compared to ceratopsians, and the mechanics of chewing differed dramatically between these two major groups. Hadrosaurs had a jaw hinge system that allowed the upper jaw to flare outward enough to allow a grinding motion that could pulverize tough stems and high food-value items such as the seeds of flowering plants. Several instances of a direct association of conifer remains with hadrosaur skeletons that could reflect stomach contents have been reported; however, none of these associations are unequivocal.[31] One thing is quite clear: hadrosaur tooth batteries and jaw mechanics were adapted to chewing tough, fibrous, and silica-rich foods. Ferns, cycads, grasses, and sphenophytes such as *Equisetites* fit the bill (pardon the pun). All of these plants had abundant representatives in the Late Cretaceous of the paleo-Arctic. These dinosaurs could also have consumed foliage and fruits of the contemporaneous and evolving flowering plants.[32]

In ceratopsids, the teeth in the upper jaw came down just outside the teeth in the lower jaw and made contact along a vertical plane. Thus, chewing in ceratopsids took the form of a scissors-like, shearing motion rather than a crushing and grinding motion.[33] However, there is a second evolutionary convergence seen in hadrosaurs and ceratopsians. Both groups evolved toothless areas in front of their tooth-lined jaws. The hadrosaurs had a flared duck-like toothless "bill" that was probably covered with a thin keratinous sheath (see figures 6.5A and 6.6). That such a sheath was

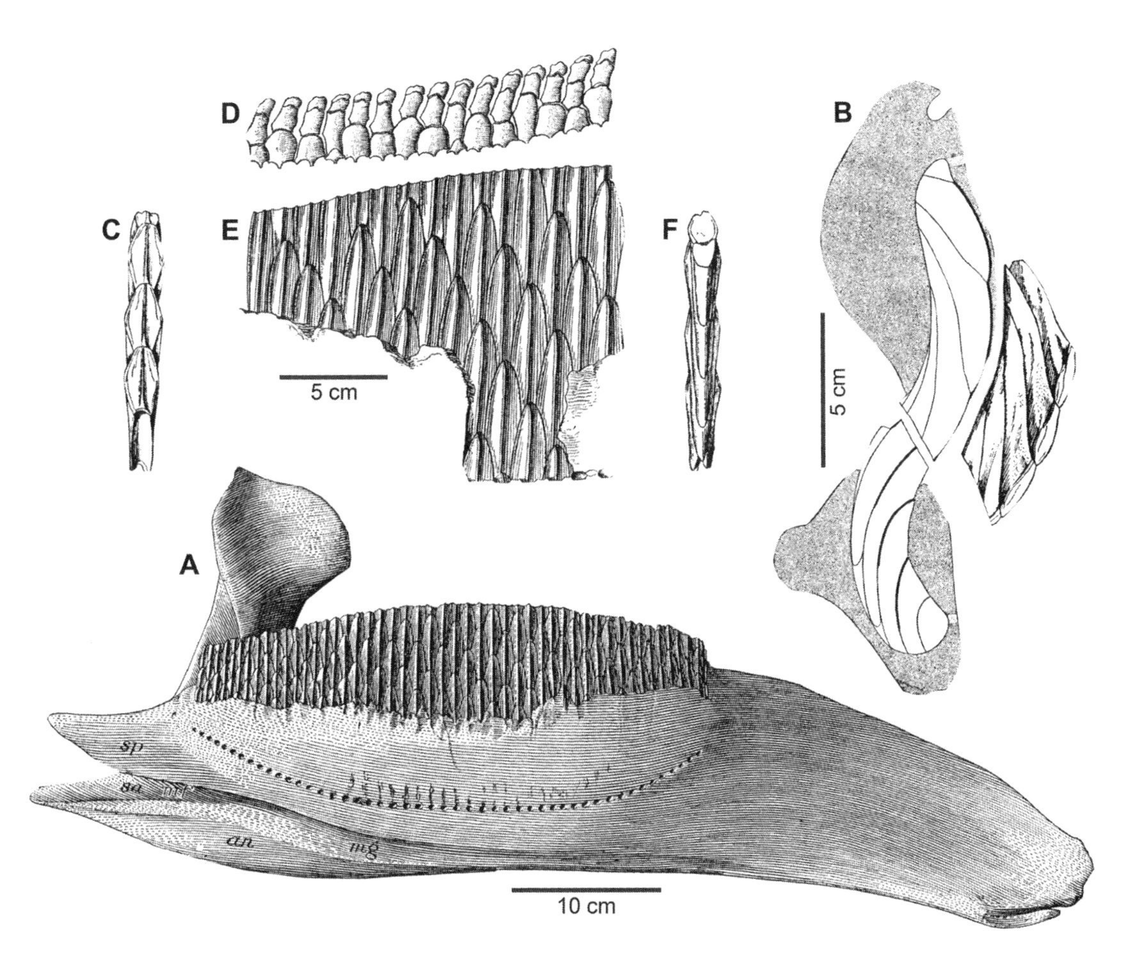

Fig. 6.5. *Drawings of Edmontosaurus* lower jaw and tooth batteries: A. Internal view of left lower jaw showing teeth in place. Scale = 10 centimeters. B. Diagrammatic cross-section of upper and lower jaws showing teeth in place and occlusion of erupted teeth. Scale = 5 centimeters. C, E: Internal view of tooth rows with erupted teeth at top. Scale = 5 centimeters. D. View from above lower jaw showing worn teeth along occlusal surface. Scale = 5 centimeters. F. External view of single tooth row from lower jaw. Scale = 5 centimeters. *Credit: Lambe (1920), Lull and Wright (1942), and David Smith.*

present is implied by the concentrations of blood vessel tracks in the bone. Ceratopsids had a set of narrow pincher-like bones that resembled the beak of a parrot that also appears to have been covered be by a keratinous sheath (see figure 6.7). When the entire jaw apparatus of each group is compared, it is clear that they differed in their feeding habits. Exactly how they differed has been the focus of a long debate among paleontologists. Perhaps the hadrosaurs were obligate grazers and "shovelers," while the ceratopsids were more like the wood bison of Alberta and obligate browsers that could nip off individual shoots or branches, as well as fruits and reproductive bodies. Therefore they were most probably not in direct competition for food but occupied different niches in the same habitats. Two things stand out vividly when the fossil records of these two herbivorous clades are reviewed. First, the fossils of some taxa of both clades are often found in bone beds within sedimentary contexts that imply very similar environments. This is especially true for the hadrosaurine taxon *Edmontosaurus* and the centrosaurine taxon *Pachyrhinosaurus*, both of which are consistently associated with wetlands. These wetlands represent lacustrine, floodplain, and coastal lowland settings.[34] These bone beds are dominated by one or the other taxon. In addition, fossils of other herbivores are either absent or very rare. This strongly implies that these taxa were gregarious and possibly social.

Fig. 6.6. Composite reconstructed skull of late juvenile *Edmontosaurus* showing right lateral view. Bones are from the Liscomb Bone Bed. *Credit: Mark Goodwin and Patricia Vickers-Rich.*

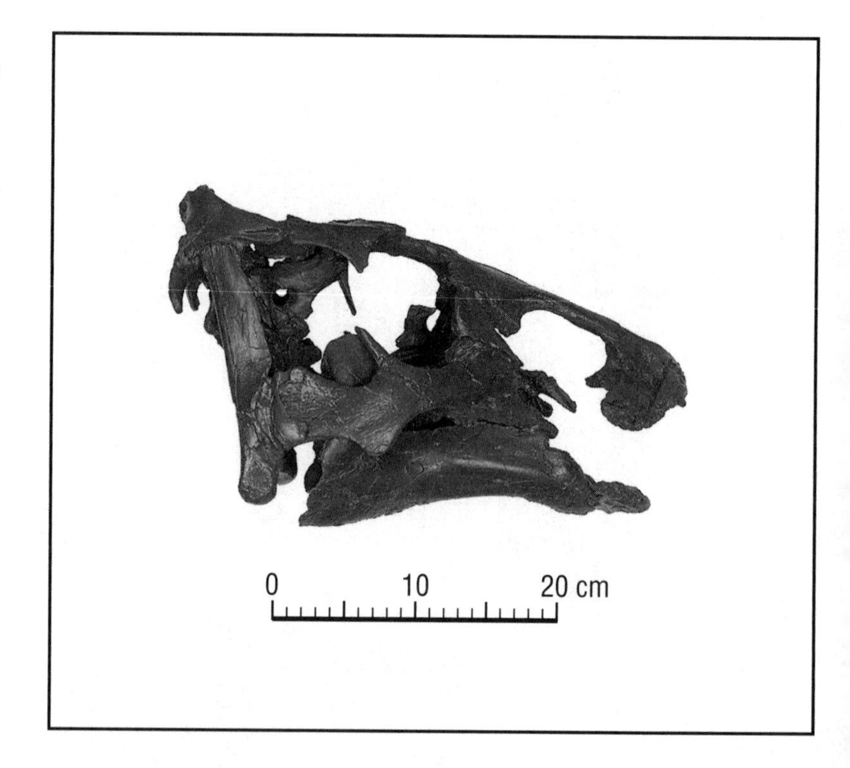

Abundant trackways that are attributable to either one or the other taxon support such a conclusion.

The ceratopsids such as *Pachyrhinosaurus* might also be found in large herds or socially interacting aggregations, but as indicated by the distribution of their remains, they would have most likely concentrated more of their feeding among the coniferous forest belts that occupied the upper margins of the deltas and the lower foothills of the emerging Brooks Range. *Pachyrhinosaurus* could have partially overlapped with the hadrosaurs along the higher and drier parts of the levee systems (see figure 6.2). The ceratopsians, with their jaw mechanics, would probably have been snipping off fronds of ferns and deciduous cycads such as *Nilssoniocladus*. In addition, they could have been stripping the leaves from stems of ginkos and the needles off a variety of deciduous conifers such as *Parataxodium* and *Podozamites*. *Pachyrhinosaurus* and close kin probably ate a greater volume and variety of vegetation with higher food value than their *Edmontosaurus* neighbors, whose grinding jaws allowed them to derive more nutrition from lower quality plants. The fruits and larger seeds of evolving flowering plants could have provided concentrated food value. Angiosperm seeds and pollen have been recovered from the Liscomb Bone Bed.

Fellow Travelers: Boneheads, "Tanks," and Other Herbivores

Represented by far fewer numbers of bones and teeth than the hadrosaurs and ceratopsids are three other lineages of herbivorous dinosaurs: the pachycephalosaurs, ankylosaurs, and basal ornithopods. The pachycephalosaurs are popularly known as dome or boneheads and are closely related to the ceratopsians. The ankylosaurs (see plate 6) are tank-like and

Fig. 6.7. Pachyrhinosaur specimen TMP 2002.76.1 from the Dinosaur Park Formation near Iddesleigh, Alberta, Canada. Original bones mount in the ceratopsian gallery of the Royal Tyrrell Museum, Drumheller, Alberta, Canada

are closely related to the stegosaurs, the typical armored dinosaurs of the Jurassic period. Basal ornithopods are a stem group that is more closely related to hadrosaurs than to any of these other clades.[35] Thus far, the paleo-Arctic pachycephalosaurs are only represented by teeth and a single skull bone. The single bone, a squamosal, was collected from a site on a slough just west of the Colville River and a few miles upriver from the Liscomb Bone Bed.[36] The pattern of nodes (bumps) and the thick bones of the pachycephalosaur skull roof, found in Late Cretaceous taxa, make this taxon very distinctive and easy to recognize, even from fragmentary fossils (see plate 7 and figure 6.8). Little is known about the behavior or feeding habits of this group of small- to medium-sized herbivores, which ranged in length from 2 to 15 feet (0.7 to 5 meters). Their teeth are quite similar to those found in ankylosaurs, making it quite hard to distinguish this taxon solely on teeth. The close similarity in tooth form, along with the shape and placement of cheek teeth, bespeaks evolutionary convergence and most probably points to a close similarity in food preferences and processing. However, these two taxa differ markedly in the details of their jaws and chewing apparatus.

Ankylosaurs have a toothless and slightly flared snout that vaguely reminds one of the spoon-like bill of hadrosaurs. Several aspects of their cranial structure suggest complex chewing motions.[37] Pachycephalosaurs, in contrast, have peg-like teeth in front of the main dental arch in the upper

A

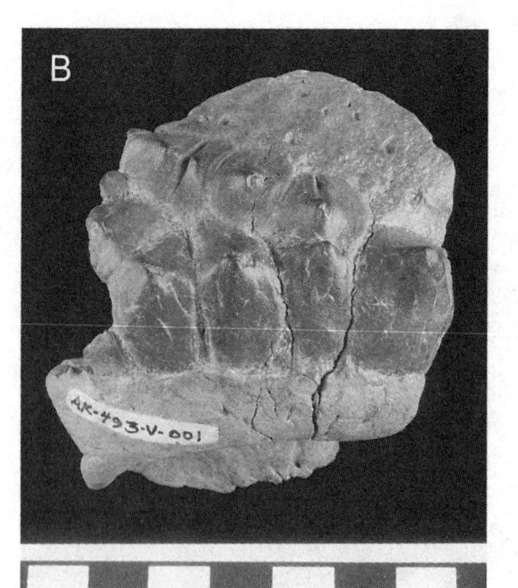
B
AK-493-V-001

C

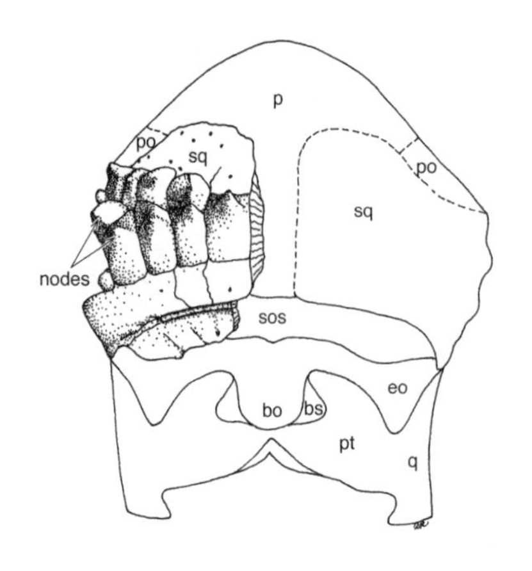
p
po
sq
po
sq
nodes
sos
eo
bo
bs
pt
q

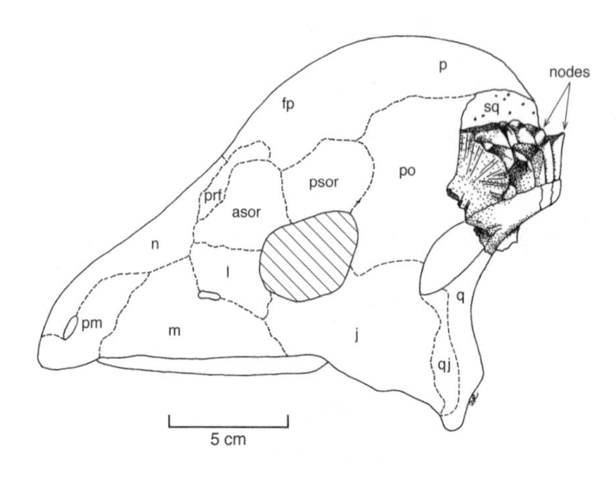
p
nodes
fp
sq
prf
psor
po
asor
n
l
q
pm
m
j
qj
5 cm

jaw but no teeth in front of the main dental arch of the lower jaw. The skull structure lacks secondary hinges, reflecting a simple mechanical, nonflaring chewing motion. Pachycephalosaurs may have been adept at snagging fruits and leaves off the branches of a variety of angiosperms along levees and in the understory of mixed forests that were found on higher ground between floodplain channels. Ankylosaurs would most likely have focused on gathering leaf litter on levees and forest floors as well as on raking in algal and rhizomal mats along lake and ocean shorelines. The paleo-Arctic ankylosaurs are found in the Late Cretaceous of Alaska and Koryakia, such as the partial skull and teeth of the nodosaurid *Edmontonia*, that were found in Late Cretaceous shallow marine deposits assigned to the Late Cretaceous (Campanian to Early Maastrichtian) member of the Matanuska Formation in the Talkeetna Mountains 93 miles (150 kilometers) northwest of Anchorage, Alaska.[38] Some of the many teeth collected on the North Slope from microvertebrate sites are probably attributable to ankylosaurs. Teeth identified as those of an ankylosaur are reported from Kakanaut in Koryakia.[39] Ankylosaur remains are consistently found in near-shore marine, alluvial floodplains and wet deltaic environments. Pachycephalosaurs are relatively rare wherever they are found. To date, they have been recovered from quite a range of wet to dry alluvial environments but exhibit a slight preference for drier conditions. No record of pachycephalosaurs has been reported from marine sediments as yet.[40]

The pachycephalosaurs and ankylosaurs are accompanied by another minor faunal element in the paleo-Arctic—basal ornithopods. These small "gazelle-like" herbivores have teeth and jaws more similar to pachycephalosaurs than to hadrosaurs, ceratopsids, or ankylosaurs. This faunal element is represented by rare teeth found on the North Slope of Alaska that were initially and tentatively attributed to an indeterminate hypsilophodont but are more likely teeth of *Thescelosaurus*.[41] Two small scapuli and several teeth found in Kakanaut have been reported as closely resembling those of *Hypsilophodon foxii*.[42]

We now have a diversity of taxa in the Late Cretaceous of the Arctic that have typically been tagged as herbivores. The evidence from fossil dung as well as the teeth and jaw mechanics of the hadrosaurs and ceratopsians strongly suggests that they were herbivorous. However, the pachycephalosaurs, ankylosaurs, and basal ornithopods had similar and distinctive teeth that are not as easily categorized as either those of an obligate herbivore or carnivore. The basic form of a compressed crown with serrated margins resembles the teeth of iguanids, a type of lizard.[43]

Some or all of these taxa may have been adapted to the range of foods that modern green iguanas consume, with an emphasis on soft algae, young shoots and leaves, snails, insects, amphibians, and carrion such as rotting fish.[44] One has to be careful about using dental morphology alone in determining a fossil animal's diet and feeding behavior. For example, crocodilians were all assigned to an obligate carnivorous habit until the discovery of the herbivorous crocodyliform taxon *Chimaerasuchus paradoxus* in the Early Cretaceous of Hubei, China.[45] When it comes to food, most living animals are highly opportunistic. It is only too human to

Fig. 6.8. Images of skull bone (squamosal) of pachycephalosaur *Alaskacephale* and drawing of complete hypothetical skull (based on *Pachycephalosaurus*) showing possible position of the squamosal bone. Scale = centimeters. *Credit: Dixon Jones, Dawn Roberts and Gangloff, Fiorillo, and Norton (2005).*

overgeneralize about our fellow creatures. The gracile basal ornithopods and pachycephalosaurs, with their similar teeth and body forms, may have competed with one another for food during the latest Cretaceous, while the heavily armored and squat nodosaurid ankylosaur *Edmontonia* could have been the dinosaurian equivalent of a turtle in some of its feeding habits.

Predators: The Top of the Food Chain

The predators of the Late Cretaceous Arctic can be divided into two theropod groups: medium to large (from 26 to 46 feet, or 8 to 15 meters long), heavily built carnivores such as *Daspletosaurus*, *Albertosaurus*, and *Gorgosaurus*, and small to medium (from 6 to 8 feet, or 2 to 2.5 meters long), gracile carnivores such as *Troodon*, *Dromaeosaurus*, and *Saurornitholestes*. Both groups are represented among fossil collections from the paleo-Arctic but almost exclusively by shed or broken teeth. Evidence of direct predatory activity is scarce. In Alaska, fewer than 1 percent of over three thousand bones bear theropod tooth marks. A range of 1 to 4 percent is typical of most collections of dinosaur bones.[46] In similar accumulations of fossil mammal bones, the incidence of bite-marked bones ranges from 13 to 38 percent and reflects the mammalian predator-scavenger habit of crushing and chewing bones for marrow content.[47] This behavior appears to have been missing from a dinosaur predator's "bag of tricks." A notable exception to this general pattern of scarcity of carnivorous dinosaur tooth marks is found in collections of the Upper Cretaceous Dinosaur Park Formation in Alberta Canada. A high of 14 percent of hadrosaur bones were tooth-marked while 5 percent of ceratopsians showed evidence of theropod bites.[48]

Troodon: The Ultimate Arctic Carnivore

Quite often, as several small teams of volunteers or students were totally absorbed in excavating and mapping part of the Liscomb Bone Bed, a shriek of glee would travel along the foot of the bluff: "Wow, a *Troodon* tooth!" *Troodon*, "wounding tooth," is a very appropriate name for this fascinating small theropod. Not only were these teeth easy to identify, but they also elicited a great deal of speculation when the proportionally large, denticles (pointed cusps) slowly came to light as the dime-sized teeth were cleared of their adhering matrix. The *Troodon* teeth found in Alaska are typically larger than those found elsewhere and represent the largest proportion of the Alaskan theropod tooth samples.[49] *Troodon* in Alaska, except for two partial brain cases and a few postcranial elements, is almost completely represented by teeth. Although this theropod typically possessed a slightly larger tooth count than its contemporary theropod brethren, the disparity in tooth recovery statistics most probably reflects this taxon's greater abundance on the paleo-Arctic landscape. This small brainy theropod is also reported in Kakanaut, just southwest of the Bering Strait (see figure 2.1).

Troodon was one of the earliest dinosaurs described from both North America and Asia. Unfortunately, Leidy published the taxon *Troodon formosus* on the basis of a single tooth in 1856.[50] For almost a hundred years, the true nature of *Troodon* was obscured by its poor record and by its being lumped in with the herbivorous pachycephalosaurs. It took the discovery

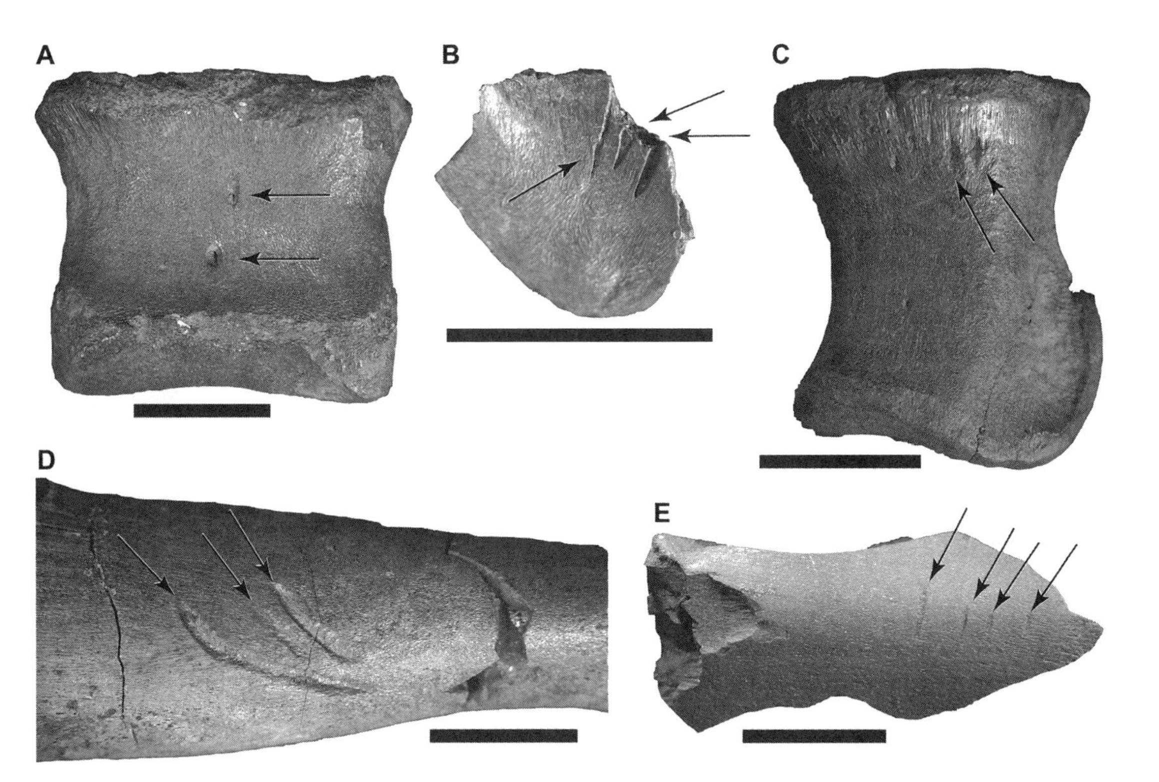

Fig. 6.9. Examples of bite marks on bones of *Edmontosaurus* from the Liscomb Bone Bed, scale = 2 centimeters. A. Two tooth pits on ventral surface of pedal phalanx 1, digit III. B. Parallel tooth scores on skull fragment. C. Two tooth pits on dorsal surface of right pedal phalanx 1, digit II. D. Three tooth scores on surface of right radius. E. Four parallel tooth scores on indeterminate limb bone fragment. Arrows indicate tooth scores. *Credit: Roland Gangloff and Anthony Fiorillo.*

of more complete material and the acumen of a group of Canadian and Chinese scientists to finally elucidate the correct nature of this highly evolved carnivore. What has now come forth is an intriguing and beguiling figure—a gracile form possessing a long slender skull that was equipped with, proportionately, the largest brain and eye of any known theropod (see plate 7 and figure 6.10).

Troodontids (*Troodon, Saurornithoides, Sinornithoides, Sinovenator,* and *Borogovia*) had brains that compare, in size, with modern flightless birds such as the emu and ostrich. The brain was equipped with large optic processing centers and a large middle ear cavity.[51] This combination most likely gave *Troodon* a highly developed light-gathering optic system that would have provided image acuity allowing for a high degree of detail discrimination.[52] This visual system, combined with great hearing acuity and good balance, would have been a dynamite toolkit for a predator that overwintered and hunted during at least seven months of continuous darkness and twilight in the Arctic. These adaptations would have given *Troodon* an advantage over contemporaneous Arctic Alaskan carnivores such as tyrannosaurids and dromaeosaurs in preying upon *Edmontosaurus*, The herbivore taxon that most likely overwintered based on the preponderance of evidence.[53] *Troodon*'s smaller size may have allowed it to resort to denning and deep sleep during unusual cold spells, something that the larger theropods could not have done.[54] Unfortunately, this preliminary hypothesis cannot be tested yet. There are, at present, none of the proper *Troodon* bones from Alaska that would allow for the proper analysis—presence or absence of lines of arrested growth.

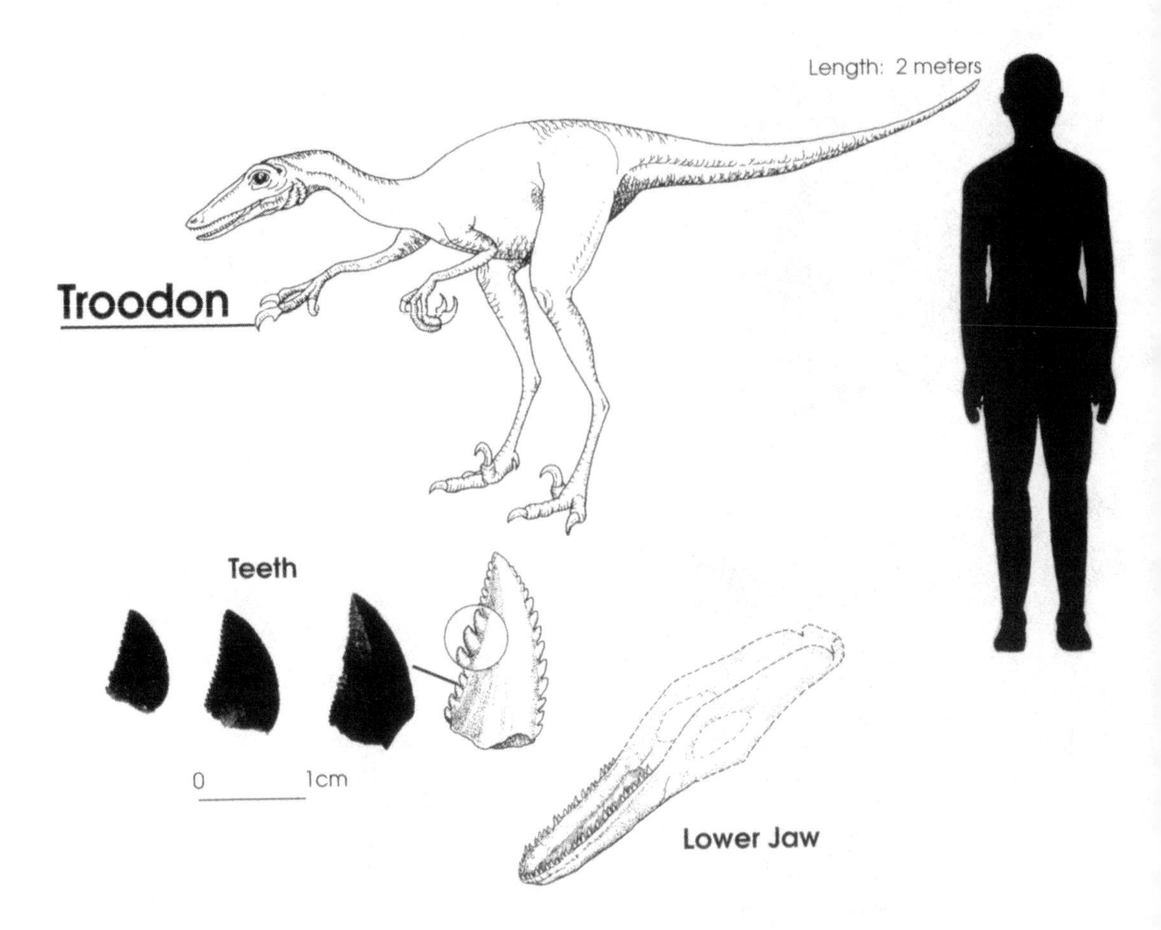

Fig. 6.10. Drawings of *Troodon* and lower jaw, combined with drawings and photos of individual teeth. *Credit: David Smith and University of Alaska Museum of the North, Fairbanks.*

What evidence points to predatory prowess in *Troodon* and the troodontids? Hollow bones, proportionally long legs and an elongate foot would have given *Troodon* the capacity for a fairly high running and pursuit speed. Add the enlarged claw on the second digit, along with the unusually large denticles (serrations) along the tooth margins, and you have a predator equipped to handle fairly large prey such as late juvenile hadrosaurs and ceratopsians, dog-sized mammals, and thescelosaurs. Quickness, a high degree of visual acuity, and the proportionally large denticles on the teeth (see figure 6.10) would also have served an insectivorous life-style. Because of their small size, *Troodon* individuals would probably not have been able to hunt subadult or adult hadrosaurs and ceratopsians. However, great visual acuity would have been a great aid in group or cooperative hunting, for it would have allowed individuals to recognize one another and would have facilitated the use of visual signals between individuals even under low light conditions. Today, coordinated group hunting strategies allow smaller mammal predators such as some felids and several types of canids to successfully hunt larger animals.

Visual signaling would have been stealthier than vocal signaling during attacks in open terrain. This and the anatomical attributes of *Troodon* might have allowed paleo-Arctic troodontids to raid *Edmontosaurus* and other herbivore nests at night. Researchers in south-central Alberta have reported an occurrence of *Troodon* and *Edmontosaurus* at a site in the

Horseshoe Canyon Formation that suggests *Troodon* was specialized to prey on the young of *Edmontosaurus*.[55] It should be pointed out that no comparable specialized predator-prey relationship has been reported for living mammal carnivores.[56] However, this does not preclude such a relationship in dinosaurs.

Dromaeosaurs and ornithomimids, based on the evidence compiled thus far, were far less abundant than *Troodon* even though they overlapped *Troodon* in size and shared several other physical attributes with it, although perhaps *Troodon*, with its proportionally larger brain and optical lobes, was capable of a more refined communication and recognition of individuals than either dromaeosaurs and ornithomimids.[57] These mental capabilities could have allowed *Troodon* individuals to engage in greater cooperation and coordination while hunting, resulting in greater success rates. It is certainly possible that if *Troodon* did develop a highly stealthy hunting strategy under low light conditions, it may also have raided the nests of other theropods. Greater hunting success when compared to other theropods combined with the ability to include theropod young on their menu would have resulted in a proportionally larger *Troodon* population. This may be reflected in the greater proportion of *Troodon* teeth recovered as noted earlier.

The abundance and variety of herbivores that have been documented for the Late Cretaceous Arctic would certainly support several tiers of carnivores. Based on relative abundance in the accumulated fossil record, it appears that *Troodon* was a secondary or tertiary consumer or predator. If indeed troodontids had developed cooperative hunting behavior, then this trait, combined with their large brain, hearing prowess, and optical advantages, would certainly have allowed this more gracile theropod to successfully compete with the larger-bodied dromaeosaurs and tyrannosaurids, and it may very well have been a higher-level predator in the paleo-Arctic while occupying a lower position in the food chain at lower latitudes.

The relatively scarce but large-bodied tyrannosaurids, such as *Albertosaurus* and *Daspletosaurus*, would have been able to prey on subadults and weakened adults of hadrosaurs and ceratopsians. In addition to having been hunting targets for the known paleo-Arctic predators, the much more abundant hadrosaurs and ceratopsians would, at times, have provided a plethora of rotting carcasses—the result of drowning, infrequent snowstorms, and epidemics. It is unreasonable to assume that such a food source would have been ignored by the theropods. It is quite possible that the much larger tyrannosaurids would have had a feeding advantage over some smaller predators because they could have chased them off carcasses after accidents (see figure 6.3) or following kills. Dominant modern predators such as East African lions (*Panthera leo*) can be observed to chase hyenas and cheetahs from their kills.[58]

The paleo-Arctic dinosaur records of Alaska and Kakanaut in the Koryak Uplands of northeastern Russia stand alone in abundance and diversity of theropods. The theropod record for the rest of the Cretaceous Arctic is greatly impoverished by comparison (see figure 2.1). However, I must remind the reader that very little prospecting has been done for dinosaur remains in most of the Arctic up to this point.

7 CRETACEOUS DINOSAUR PATHWAYS IN THE PALEO-ARCTIC AND ALONG THE WESTERN INTERIOR SEAWAY

The Association of Dinosaur Trackways with Shorelines in Western North America

The record of Cretaceous age dinosaur trackways and their common association with ocean shorelines has been steadily accumulating since the 1940s.[1] One of the most impressive and best-known records can be found in central Texas, an area that occupied the southern end of the great Western Interior Seaway. The Early Cretaceous (Albian) age Glen Rose Formation here contains numerous and widespread trackway complexes at several stratigraphic levels. Such widespread trackway complexes are now referred to as "megatracksite complexes."[2] The Glen Rose Formation includes fine examples of megatracksites that crop out over an area of some 38,000 square miles (100,000 square kilometers), including the famous Paluxy River site that was studied by the early dinosaur "tracker" Roland T. Bird, and is now a Texas state park.[3]

Sauropod and theropod tracks and trackways generally follow along Early Cretaceous marine shorelines. This is a pattern that is typical of Cretaceous-age tracksites.[4] The recent documentation of abundant tracks and trackways in Upper Cretaceous rocks of the Kaskapau Formation in northeastern British Columbia is especially relevant to a discussion of paleo-Arctic dinosaurs and their association with ocean shoreline environments. This locality is on Quality Creek near Tumbler Ridge, British Columbia. It was located above the paleo-Arctic Circle at a high stand of global sea level. Three different coastal environments are recorded in a wedge of nonmarine sedimentary rocks that is part of a marine-dominated offshore formation. The area underwent a change from a strandplain with beach ridges and sandy coals to tidal channels then to a backshore freshwater lake and finally to a brackish lagoon with a complex of deltas. In two of these environments, abundant tracks and trackways are found in localized concentrations characterized as trampled or dinoturbated horizons (see figure 4.14). In contrast with North Slope Late Cretaceous dinosaurs, the Quality Creek dinosaurs have been directly associated with crocodilians, turtles, and oyster shells. It is interesting to note that a dinosaur-trampled oyster bank, or bioherm, was found in the brackish lagoon. This trampled bioherm is a first for British Columbia, as is the discovery of in situ dinosaur bones.[5]

If you look carefully enough at the bedding surfaces of Cretaceous rocks that include coal beds you are bound to find dinosaur and bird tracks. This statement has become a mantra for dinosaur "trackers" that scour the outcrops of Cretaceous terrestrial rocks throughout western North America. The environmental conditions responsible for the accumulation of peat-rich sediments that are the first step in the formation of coal are also conducive to the preservation of dinosaur and bird tracks. The ideal conditions are water-saturated mud and fine silty sand surfaces that harbor abundant microbes such as cyanobacteria and green algae.[6] If these conditions are accompanied by influxes of fine-grained, rapidly accumulating sediments, then the making of clear impressions and their subsequent burial and preservation are greatly aided. This combination of factors describes those encountered along most delta, lake, and ocean shorelines. These habitats also contain rich assortments of plants, such as grasses, scouring rushes, and ferns. During the Cretaceous, these plants were often joined by ginkgos, cycads, and a variety of needle-bearing gymnosperms—a plant smorgasbord that would attract herbivorous duckbills, ceratopsids, pachycephalosaurs, and ankylosaurs. It is now very clear, from several lines of evidence, including abundant trackways, that Cretaceous dinosaurs lived and flourished along paleo-Arctic lake, delta, and ocean shorelines.

Arctic Dinosaur Pathways: The Great Western Cordilleran "Highway"

As noted in chapter 2, the first documentation of dinosaurs in the Arctic was the discovery of their footprints on the Island of Spitzbergen. Likewise, early evidence of dinosaurs in the Arctic of Alaska came from their footprints.[7] However, the diversity and great abundance of this record along the Colville River on Alaska's North Slope remained elusive despite over forty years of exploration by hundreds of field geologists and biologists that criss-crossed this area. This points up the value of having the "eye" for this type of fossil and the importance of field training that helps to acquire it. It is also critical to keep the mind open to paradigms other than those acquired in early schooling. Apparently, the rich record of dinosaur footprints and trackways was noticed, but the footprints were misidentified as abiogenic sedimentary structures instead of fossil tracks. Abundant and widespread Cretaceous dinosaur trackways from northern North America were first reported in 1979.[8] This study dealt with finds in northeastern British Columbia and northwestern Alberta, Canada, both part of the Western Cordillera, a mountainous region that extends from Southern California to northern Alaska.[9] It includes a complex of mountain ranges such as the Sierra Nevada, the Cascades, the Rocky Mountains, and Brooks Range, along with related plateaus, intermontane basins, and valleys. The part of the Western Cordillera that includes abundant mid- to Late Cretaceous trackways and megatracksites and connects the paleo-Arctic and subarctic latitudes (see figure 6.1) is what I call a "highway." Lockley prefers the label "freeway," but that's a metaphor that has implications that highway, likewise a metaphor, doesn't.[10]

I introduce the discussion of the Western Cordilleran dinosaur highway by recounting a personal odyssey that not only made me aware of the extraordinary Canadian record of trackways but also put me in touch with some inspiring Canadian colleagues. This odyssey begins in western Alberta at the eastern base of the magnificent Rocky Mountains.

The Smoky River Site

In the summer of 1999, Kevin May and I met our colleague, Richard McCrea, in the out-of-the-way small town of Grande Cache, Alberta. Tucked away in the beautiful foothills bordering the eastern flank of the Canadian Rockies, this mining community of nearly four thousand souls lies on a little-known shortcut to Grande Prairie and the famous Alaska Highway. We were there to learn more about the spectacular dinosaur trackways that Rich had been mapping and analyzing for two years. I had learned of his research in the Smoky River coal mining area at an annual meeting of the Society of Vertebrate Paleontology in Utah the previous year. Rich had presented his findings, complete with excellent photos and drawings, but as we were to find out, it was impossible to grasp the scale and complexity of the trackways despite the impressive visuals.

As we drove some 12 miles (20 kilometers) east of Grande Cache, we passed through beautiful mixed coniferous forests. There was little hint of the massive mining operation nearby as we followed the west bank of the Smoky River. We turned westward and followed a wide dirt road to the operational headquarters of the Smoky River Coal Ltd. We waited outside the fence as Rich went in to find his contact and to clear us for entry. The size of the complex and the number of large hauling vehicles dwarfed similar facilities near Fairbanks that I was familiar with. This began to give us some inkling of the scale of the sites that we would soon encounter.

After meeting with our guide and passing through security, we drove a mile along the main dirt access road to one of McCrea's principal study sites. This site, known as the W-3 mine, also held the largest concentration of trackways discovered at Smoky River up to that time.[11] As we stood by the side of the road, Rich gave us an overview of his work and recounted the geologic context of this immense exposure of rock that had originally underlain the thick coal seams.

The thin, shiny, dark gray mudstone covered a silty, fine grain sandstone that was part of the number 4 coal seam that dates to the Early Cretaceous (Albian) age and that is the Grande Cache member of the famous Gates Formation. The massive exposure was close to 100 feet (30 meters) high and over 0.25 miles (600 meters) long, forming the eastern flank of a large up-fold (anticline). It dipped toward us at an average angle of 60°. It was late morning when we arrived, and the sun's rays were low enough to reveal a system of trackways that were made by three- and four-toed individuals. It was all we could do to contain our excitement at the expanse of what we were seeing. Soon we were scrambling over and around large blocks of rock and mining tailings that still held pockets of winter snow. We arrived breathless and awestruck by what was within reach—that is, if you had climbing ropes, or the feet of a gecko. Series after series of three- and four-toed dinosaur tracks seemed to be going in every direction.

Initially, Rich had mapped twelve hundred vertebrate footprints in an area that measured 2,000 square feet (186 square meters). This grew to six thousand footprints, all of which he documented while hanging from his climbing harness and secured by ropes. (We suspected that he must have developed sucker pads.) As is typical, what we were now seeing close-up greatly surpassed what we thought we had grasped from the road and the

Fig. 7.1. Closeup view of negative *Tetrapodosaurus* trackway on steep footwall at W-3 mine site, Smoky River Coal Mine, Alberta, Canada. *Tetrapodosaurus* footprints cross from center-left to right edge and are 14–18 inches (35–45 centimeters) long. Note three-toed theropod trackway trending downward near center and scattered ripple-marked beds. *Credit: Richard McCrea and McCrea, Lockley, and Meyer (2001), modified by Roland Gangloff.*

images in Rich's presentation. In the morning light, we could make out three long series of small manus and larger pes imprints, two of which crossed one another, and an almost vertical series of three-toed footprints. This was our first encounter with what is known as a megatracksite.[12] Having been confined to the limited exposures of trackways afforded along the Colville River, we were not prepared to grasp what we were now confronted with. On the Colville, we had been overjoyed whenever we discovered trackways covering a few tens of square feet, but what we were attempting to take in here at the Smoky River mines boggled our minds.

By the end of two days, we had explored six other tracksites and had seen a trackway assemblage that was dominated by the four-toed ichnogenus *Tetrapodosaurus*.[13] This ichnogenus is most likely attributable to armored dinosaurs (see plate 6). At the W-3 site alone, Rich had mapped over twenty trackways, some with over fifty consecutive footprints, assignable to this taxon. In addition, at least five other ichnotaxa attributable to nonavian theropods and birds had been seen at the six track sites that we had visited. Within a mile of the W-3 site we came across a site with such concentrations of footprints on the bedding surfaces that the label "dinoturbated" seemed quite appropriate. At another site, we saw upright fossilized tree trunks with trackways skirting the splays of the tree roots. Several sites held wave-rippled (see figure 7.1) and churned-up surfaces with trackways that reflected dinosaurs walking along lake shorelines or perhaps tidal mudflats. In total, Rich mapped over sixteen sites contained within some 16 square miles (41 square kilometers), yielding over six thousand footprints representing dinosaurs, birds, and an early mammal.[14] We all came to the conclusion that if coal-mining were ever discontinued, this area should be preserved as a national or world geologic heritage site. In 1999, Smoky River Coal Ltd. abandoned operations, but the properties were taken over by Grande Cache Coal Corporation that has expanded operations. The good news is that continued mining will expose more trackways. It is important to note that several of the world's most famous and productive fossil sites, such as the Solnhofen in Bavaria, were exposed and furthered by mining.

The Whiskers Lake-Ross River Site

On the way back to Fairbanks, Kevin and I decided to take a little side trip to Ross River, 143 miles (230 kilometers) northeast of Whitehorse in the Yukon Territory of Canada (see figure 1.2). We had heard that it was a beautiful area and was not far off the path from our intended destination of Dawson City; besides we needed a rest from driving and it was near lunchtime. This proved to be a very fortuitous decision. As we passed roadcuts along the northern edge of Whiskers Lake just outside the small First Nation settlement of Ross River, Kevin and I noticed indications of coal mining operations to the north of the road, along with some suspicious patterns on exposed bedding surfaces. Having just come from Smoky River, and with our minds flooded with images of trackways, we couldn't resist taking a closer look. Geologists and paleontologists are notorious for not keeping their eyes and minds on the road while driving past well-exposed rock outcrops. Noticing the pattern of rocks and constantly processing these images becomes part of how geologists see the world. Within fifteen minutes of exiting the truck, Kevin's "They're here!" announced the discovery of dinosaur footprints. Within an hour, we had cleared enough of the talus and vegetation to determine that we had a system of trackways, not just a single footprint. A further drive up an old mining road led us to a fairly large abandoned open-pit mine whose southern end was filled with a deep pond. The whole western flank of the mining pit was made up of a steeply dipping rocky slope similar to the slopes at Smoky River. We didn't have the time to properly prospect the area, but we vowed to return.

We contacted our colleague John Storer in Whitehorse to determine what was known about this site and whether we could gain legal access. John was the official paleontologist for the government of the Yukon Territory and a specialist in vertebrate paleontology. Within a week we had John's excited response. He declared that Kevin and I had discovered the first dinosaur trackways in the territory and that the rocks yielding the trackways had been assigned an Eocene age based on fossil pollen found in similar rocks in the area. This meant that we could possibly have discovered the first post-Mesozoic dinosaur fossils or, more likely, that the Eocene-age assignment was incorrect. In the backs of our minds, we hoped for the former but realistically expected the latter, as the track- and coal-bearing sedimentary sequence sits within a complex system of fault blocks. This made the determinations of geologic age relationships very difficult.

Within a year, a team from our museum, working with counterparts from the government of the Yukon Territory, had confirmed and then fully evaluated and documented the area directly related to the discovery site. Funding from several Yukon territorial agencies and the Yukon Institute supported the fieldwork. Footprints and/or trackways were discovered at two other stratigraphic horizons within a rock package totaling over 1,200 feet

Fig. 7.2. Dense *Tetrapodosaurus* trackways and associated tree stumps with radiating roots on steep footwall at E-2 coal pit, Smoky River Coal Mine. Scale = 1 meter. *Credit: Philip Currie and Royal Tyrrell Museum of Palaeontology and Alberta Community Development.*

Fig. 7.3. Trackway along edge of road at Whiskers Lake near Ross River, Yukon Territory, Canada. See accompanying interpretive map for symbols and scale. *Credit: Gangloff, May, and Storer (2004), modified by Gangloff and David Smith.*

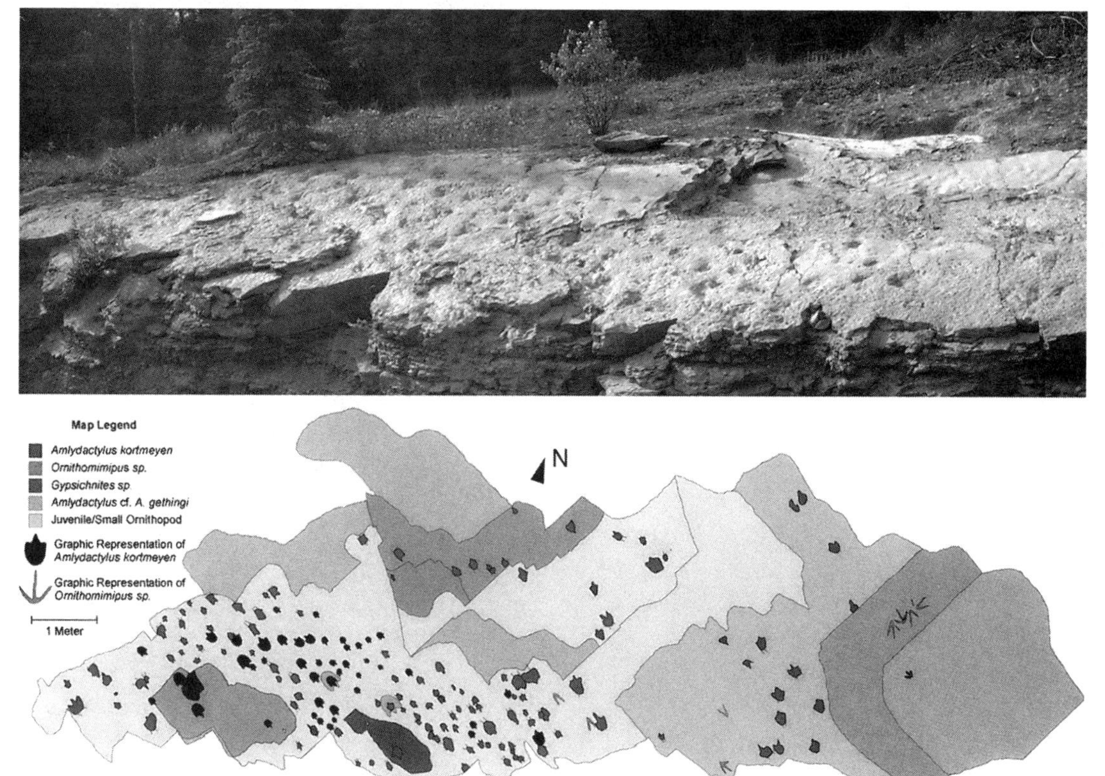

(360 meters) of strata. A total of 251 footprints were mapped or collected, and a densely trampled surface with no clear footprints was cleaned and recorded. The ichnofauna, or ichnite assemblage, is dominated by *Amblydactylus*, an ichnotaxon that is usually attributed to hadrosaurs. Five other ichnogenera attributed to theropods were identified and described.[15]

Alas, we were not able to claim to have discovered the first record of post-Mesozoic nonavian dinosaurs. Our discovery and subsequent efforts at

Fig. 7.4. Five-year-old Y. O. Kelly May with natural cast of a three-toed theropod footprint attributed to *Irenesauripus* on talus block near his right arm at Nadahini Coal Mine pit near Ross River, Yukon Territory, Canada. *Credit: Roland Gangloff.*

Whiskers Lake and the inactive Nadahini Coal Mine spurred a restudy of the stratigraphy and palynology that led to a new age assignment of Aptian to Cenomanian. This age is well within the Cretaceous and a time when the sediments of the Nanushuk Formation were a dinosaur "stomping ground" in the paleo-Arctic of Alaska. By the time the field season was over, we had laid the foundation for further work, and it was clear that much remained to be uncovered. The possibility that the main roadside site was potentially a megatrack locale made us cautious about further excavation until the territory's government and the local First Nation peoples, the Dena, could agree on the future of the site. Some individual ichnites and the silicone rubber peels of the main trackway are maintained in the collections of the Office of the Yukon Paleontologist in Whitehorse.

The Peace River Region

The Peace River originates in British Columbia's Rocky Mountains and flows east to Lake Athabasca. The river's name signifies a peace treaty between the Beaver and Cree peoples that was confirmed by the smoking of a traditional "peace pipe" in the 1790s near Lake Athabasca. The Peace River has entrenched a sinuous course through the plains of Alberta and eventually joins the McKenzie River on its way to the Arctic Ocean. The river has exposed beautiful bluffs composed of hundreds of feet of Lower to middle Cretaceous sedimentary rocks. These continental sediments are part of a thick sequence (up to 6,000 feet [1,830 meters]) of sandstones, siltstones, and conglomerates that are capped and interbedded with mudstones and shales of marine origin. Most of these rocks lie below the surface and were once part of huge complex deltas that were splaying eastward into the shallow Western Interior Seaway.[16] The porous sands, conglomerates, and interbedded coals harbor significant amounts of oil and natural gas. This ancient landscape and geologic chemistry have much in common with Alaska's North Slope. The younger package (Cenomanian) of sediments contains some dinosaur trackways, but it is the older (Aptian-Albian) package that has produced the largest number of dinosaur and bird footprints.[17]

The first footprints from this older Gething Formation were noted by a field geologist named F. H. McLearn in 1922. In the 1930s, the famous Canadian paleontologist Charles Sternberg led the first expeditions that documented and collected the dinosaur footprints and trackways, compiling a record of four hundred footprints. In the late 1970s, Philip Currie of the Royal Tyrrell Museum of Palaeontology documented some seventeen hundred footprints, including over a hundred trackways. Currie and his Royal Tyrrell crews did a tremendous job of documenting and salvaging dinosaur footprints and trackways, along a stretch of the Peace River between the upper dam and the proposed site of a new Peace Canyon Dam. Though the resulting Dinosaur Lake submerged the trackways, sections of a salvaged trackway are on display at the Peace Canyon Dam's visitor center. The great majority of the footprints and trackways are assigned to the ichnogenus *Amblydactylus* that is usually attributed to hadrosaurs but could represent other ornithopods.[18]

When the aerial extent of the trackway-bearing rock units that are found in the Peace River region are taken into account, it is clear that this area deserves to be labeled as a "megatracksite."

Evidence of a Dinosaur Highway

When all of the previously described trackway sites are compared with respect to their ichnite assemblages, geologic age, and paleoenvironmental context, an intriguing picture begins to take shape. It suggests an Early to Late Cretaceous (Aptian-Turonian) dinosaur migratory route that spans some twenty-five to thirty million years, which might be characterized as a long-term migratory "highway."[19] Dinosaur track and trackway sites often fill in where dinosaur bones (body fossils) are lacking. Therefore, they give a more complete paleogeographic picture than body fossils alone.

The Western Cordilleran long-term migratory highway (see figure 6.1) appears to have run along the western shoreline of the Western Interior Seaway, from what is now northwestern Alaska to at least central Alberta—a distance of nearly 2,000 miles (3,200 kilometers). The pathway may have split into a shoreline route to the east and a more interior route to the west. Additional finds in the Dakota Group of Colorado and New Mexico as well as the Cedar Mountain Formation of Utah may extend the age and lead to a clearer picture of the southern part of this hypothetical Cretaceous dinosaur highway.[20]

Another aspect of this period in the history of the Western Cordillera also comes into sharper focus when the dinosaur faunal records are taken into account. Hadrosaurs, ceratopsians, and ankylosaurs were widespread in western North America from the end of the Early Cretaceous to the end of Late Cretaceous and appear to have been the most frequent travelers along the proposed route. An ichnofauna comprised of *Amblydactylus, Tetrapodosaurus, Irenesauripus, Columbosauripus,* and *Gypsichnites* has now been reported from Arctic Alaska, the Yukon Territory, and northwestern Alberta. These data indicate that during the mid- to Late Cretaceous, there was a large area that shared closely similar dinosaur faunas in the paleo-Arctic of western North America.[21] Furthermore, a comparison of the body fossil record from Australia and Eurasia and from Alaska to Mexico strongly suggests that the most widespread lineages (hadrosaurs, ceratopsids, ankylosaurs, tyrannosaurs, dromaeosaurs and troodontids) of Cretaceous dinosaurs that are found in North America had their origins in the Old World.[22] What intercontinental route or pathways did these dinosaurs take to get to North America, and can we envision this route, or routes, as long-term migratory "highways"?

The abundant evidence in support of a Western Cordilleran migratory route suggests that this pathway would have served as a long-term intracontinental dispersal route that witnessed thousands of gradual migrations over a variety of distances, ultimately spreading populations of dinosaurs primarily north to south over a period of at least twenty-five million years. It is important to note that we have no evidence of megatracksites or trackways that are continuous all along the western North American pathway and that were formed at exactly the same time. Remember that "highway" as used here is a metaphor.

The Alaska-Asian Connection

If you are a Yupik or Siberian Chukchi Eskimo, you have no difficulty with the idea that Asia and North America have been physically one during the past. There are two desolate, unequal, flat-topped hunks of volcanic rock separated by a couple of miles of open water and perched on a submerged platform that lies some 98 to 160 feet (27 to 50 meters) below the present ocean surface. Before 1990, these two islands were on opposite sides of a political "Ice Curtain."[23] These two barren dots in the Bering Strait were the closest geographic points between the Soviet Union and the United States during the Cold War. A great deal of research has focused on the shallow platform that supports these islands and extends for hundreds of miles north-south and east-west of these islands. Scientists have amassed a mountain of evidence, both geologic and biologic, that supports the theory that before and after the time the Late Cretaceous dinosaurs arrived, North America was populated by a series of migrations that crossed what is now the Bering Strait from Eurasia. The close affinities of a wide variety of organisms that are shared by Eurasia and North America today are explained by this theory. Most of this research has focused on part of the last major glacial period and the subsequent interglacial warming (forty thousand to seven thousand years ago). Known to most as the Bering Land Bridge, this intermittent subaerial connection reached its greatest development some eighteen thousand years ago. This was a time of the lowest stand of the oceans, and coincided with the height of the last major period of continental glaciation. Although labeled a land bridge, this area should not be visualized as a narrow land crossing. Designated Beringia (see figure 6.1) by leading researchers, it was a subcontinent-sized migration route and system of habitats that was complex enough to generate genetic variation and speciation within the populations of organisms that occupied and traversed it.[24]

Research conducted over the last thirty years in Eastern Siberia, Texas, Alaska, the Canadian Arctic, and Alberta, Canada on Cretaceous dinosaurs, combined with the most recent plate-tectonic data has resulted in a hypothesis that proposes the existence of a Cretaceous western paleo-Arctic Eurasian-North American land connection.[25] This earlier Bering Strait bridging connection is now also referred to as Beringia. This working hypothesis is well supported when the Late Cretaceous dinosaur fauna of Alaska is compared with the fauna from Kakanaut, Russia, the Horseshoe Canyon Formation of Alberta's Red Deer River valley, and the Wapiti Formation of Grand Prairie, Alberta, Canada. The presence of genera such as *Edmontosaurus*, *Pachyrhinosaurus*, and *Edmontonia* represent clades that originated in Eurasia and are now present in Alaska's and Alberta's faunas. The presence of *Troodon formosus*, *Dromaeosaurus albertensis*, and *Saurornitholestes langstoni* in Alaska, Alberta, Canada as well as teeth from Kakanaut that may be cospecific also support such a conclusion.[26] The possibility of Cretaceous dinosaur migrations into North America from Asia have also been pointed to by researchers in the Amur region of far-eastern Russia and in Utah.[27] A Cretaceous Beringia is also supported by mammal, amphibian, freshwater fish, and insect faunas in Asia and North America.[28]

The preponderance of evidence supports a conclusion that the primary direction of these faunal exchanges between Asia and North America were from west to east. Opposite movements may be reflected in the initial migration of Asian hadrosaurid stock into North America in the Campanian and diversification of corythosaur and parasaurloph lambeosaurines with subsequent migrations of evolved representatives of these stocks back to Asia during the Maastrichtian.[29]

Was There Another Major Migratory Pathway during the Cretaceous?

Another possible long-term migratory pathway has been proposed for the introduction and spread of Cretaceous dinosaurs and contemporaneous floral and faunal elements into North America from Eurasia. This pathway has been named Barentsia and is a route that crosses northern Asia and then enters North America by way of northern Europe and then Greenland (see figure 6.1).[30] There is less evidence to support the existence of this intercontinental migratory pathway than there is for Beringia. However, the dinosaur record of the eastern Arctic is far smaller in numbers of fossils, localities, and taxonomic diversity than that of the western Arctic. Therefore, one should keep an open mind concerning the possibility of this other paleo-Arctic route.

The last three decades of research in the Arctic has unquestionably established this region as a rich home ground for diverse and thriving dinosaur populations during the Cretaceous. Our knowledge of the adaptive strategies that these animals exercised in these paleo-Arctic environments is still sketchy and far from adequate. Much more research needs to be focused on paleoecologic aspects such as winter food sources and the possibility of hibernation, as well as denning as a strategy for winter survival.[31]

The discovery and study of dinosaur track and trackway sites in the Arctic and elsewhere in Alaska are in an embryonic stage. There are a series of widely scattered track and trackway sites on the Alaskan Peninsula and in the Nanushuk Formation far to the west of the Colville and Awuna rivers (see figures 2.1 and 3.1) in northern Alaska.[32] Preliminary discoveries and the mapped extent of the Nanushuk Formation indicate that the northwestern part of the North Slope may ultimately prove to be a megatrack site. The data from all of these Alaskan sites are preliminary, and the sites beg for further evaluation and documentation. One thing is quite clear. Alaska and the Arctic still hold many keys to our understanding of the dynamics and timing of the great exchange of terrestrial Mesozoic life that took place between the Old and New World. Alaska, without a doubt, holds a strategic position in Mesozoic biogeography and will be the location of many more significant discoveries.

This was the ancient North American stage on which Arctic Alaskan and Canadian dinosaurs played out their lives and evolutionary histories during the Cretaceous. The next three chapters take a look at the impact of new technologies and economic development on dinosaur research as well as look at the possible future of research on dinosaurs in the Arctic.

APPLYING NEW TECHNOLOGIES TO THE ANCIENT PAST

8

Satellites, Lasers, and X-rays

Over the last three decades new techniques and technologies spawned by revolutions in molecular biology, computers, and microelectronics have had a great impact on paleontology and paleobiology. During my eighteen years of fieldwork on the North Slope and Arctic Coastal Plain of Alaska, I went from relying on topographic maps published by the U.S. Geological Survey and a magnetic Brunton compass to taking advantage of a series of increasingly more sophisticated and higher precision hand-held GPS instruments.

As I completed my last fieldwork on Alaska's North Slope in 2005, GPS was just beginning to supplant the use of staked grids, tape measures, and plumb bobs that had been the mainstay of dinosaur excavation and mapping for some sixty years. This satellite-based system evolved from a single GPS satellite in 1978 to a system of twenty-four satellites that encompasses the Earth from pole to pole. GPS was a tremendous aid to me in accurately locating positions on the tundra flats and river floodplains using a small handheld unit. As satellite signal security protocols were lifted by U.S. intelligence and military authorities in the late 1990s, accuracy in handheld systems went from tens of yards to less than a yard and then to a matter of inches using a multichannel total station setup.

Total stations presently exceed the accuracy and precision of the best nonelectronically based standard methods when custom computer algorithms are added. However, it will take some time to bring this more expensive and technically demanding methodology and instrumentation within the reach of most paleontological excavators. The elimination of magnetically influenced instruments is especially important for precise surveying and positioning throughout the high Arctic, where the Earth's magnetic field converges near but not at the geographic pole. By the time of my last expedition, all of our aircraft were fully reliant on GPS to navigate the vast distances and almost featureless landscapes (especially during the long winters) of parts of the Arctic. The future will bring even greater accuracy and precision as ranging lasers are integrated into the whole system. The mapping of excavations in 3-D will soon reach a precision of less than ½ inch (1 centimeter) to four plus places beyond the decimal point.

The laser was born in the Bell Laboratories in the late 1950s.[1] As a graduate student in Berkeley, I recall reading about these intense beams of coherent light in Professor Towne's University of California, Berkeley, labs in 1967. At the time it seemed to be another fantastic electronic gizmo, just like the transistor. I had no idea that it would impact not only my career but the quality of modern life. The laser continues to find important applications in paleontology and paleobiology beyond use as a field-surveying instrument. It has become the key to effective, noncontact, imaging, as these instruments can construct a highly precise, digital surface "map" of a bone or tooth. When a bone is scanned in all three dimensions, the images can be assembled using a computer. The digital files can then be attached to a modeling machine that can duplicate the bone out of a range of materials, providing precise replicas that can be sent to other researchers or become a part of an exhibit in a museum. These laser scanners are relatively inexpensive and easy to use on a variety of objects from tiny mammal teeth a hundredth of an inch (a few millimeters) long to an entire human body.

I began a program in the late 1990s to develop digital images of the bones and teeth that were being integrated into the University of Alaska Museum's Earth Science Department's collections. The goal was to reconstruct a representative, virtual 3-D map of the Liscomb Bone Bed that would join the North Slope dinosaur exhibit in the museum's public area. Visitors would stand before an electronic screen, don a special helmet, and view a 3-D image of the Liscomb Bone Bed. In addition, the process of construction of the digital graphic files would require a focus on details that would aid research. The process would also provide an opportunity to learn how to use a powerful new tool that had recently been developed on campus—a high-powered parallel supercomputer. I secured the help of my lab assistant and an undergraduate engineering student, and the initial results were very encouraging, but I was unable to complete the project. I hope that my successor will be able to take up where I left off. Not only will the planned exhibit be worthwhile, but the availability of a set of highly detailed digital images of the most significant bones in our collection will allow other researchers to study the bones without having to travel to Fairbanks. This will both make the dinosaur and Pleistocene mammal collections from Alaska more accessible and make them much more valuable for research. These scanners can only image the surface of bones and teeth. It takes other methods to see below the surface.

X-ray images have been used since the 1950s to study fossil bones and teeth using a noncontact approach. The simple dental or larger X-ray instruments have now morphed into large complex machines that can take a series of one-dimensional film slices called tomographs. First used in medical studies over thirty years ago, the last fifteen years have witnessed the application of this technique to paleontological studies. CAT scans, using complex computer algorithms, can produce 3-D volumetric images of fossils for research and museum exhibits.[2] Hundreds of studies each year rely on extensive use of this technique to study the contents of dinosaur and fossil bird eggs as well as the brain cavities of a wide variety of avian and nonavian dinosaurs. The intricacies of vertebrae, the developing teeth in jaws, and the labyrinthine inner ear of ceratopsids can now be reproduced in digital files and translated into images for study or reproduced as plastic models.

Just before I retired, I took advantage of the generosity of some of the local hospital staff and had them subject a dinosaur egg specimen brought to me from China by a student to a CAT scan. The scan did not reveal a developing embryo, much to our disappointment. However, it did allow us to get some needed practice applying this technique.

Recently, the work of Lawrence Witmer and Ryan Ridgely has revealed details of the brain cavity and inner ear of *Pachyrhinosaurus lakustai* from the Wapiti Formation of Alberta, Canada.[3] This newly defined northern pachyrhinosaur appears to be closely related to the *Pachyrhinosaur* specimens found on Alaska's North Slope. The lack of even a fairly complete specimen from Alaska makes comparison on gross anatomy alone problematic. However, the CAT files generated by Witmer and Ridgely will be a valuable aid in detailed comparisons that are now underway, since the brain and inner ear cavities could be present in at least one of the specimens from Alaska (see figure 5.8).

The last decade has seen the development of high-resolution CAT and nano-CAT with resolution at the scale of one billionth of a meter. In addition, X-rays have been augmented by other high-energy sources, such as gamma rays, that are proving to be even more revealing when applied to fossil specimens. If the trends in instrumentation, especially nano-CAT technology, that are already established continue, paleontologists in the near future will have whole new ways of studying the internal details of dinosaur remains from the Arctic.

New Instruments and New Chemistry Greatly Improve Fossil Preparation

Following collection in the field, specimens almost always require preparation before they can be studied. Fossil preparators have been especially keen to adapt any technology that would speed up and refine the process of cleaning fossil surfaces of adhering rock matrix. The use of microabrasive units that were originally developed for electronic and computer industries has truly revolutionized the fine-scale preparation of fossils. These compressed air-powered, pencil-sized "sandblasters" use a variety of abrasives from soft gypsum to diamond and carbide dust to clean fossil surfaces and reveal highly delicate spines and surface ornamentation that were often lost using earlier mechanical means such as fine chisels and dental picks. Microabrasive units are especially effective when the rock matrix is cemented to the fossil's surface by silica. When carbonate rock cements are encountered, acids can be used to dissolve the cements, but these often dissolve fossil surfaces as well. The microabrasive units greatly speed up surface preparation of a great variety of fossils.

Fossil preparators are always looking for more effective ways of consolidating and gluing together fossil remains. During the last twenty years, new forms of consolidants and adhesives, from epoxies to cyanoacrylates (Crazy Glue) have appeared. These chemicals greatly aid in the process of extracting specimens in the field and then putting the bone and tooth "jigsaw" puzzles back together. Many a fragile bone or tooth that would have been trashed during excavation twenty years ago has been saved with a very low-viscosity cyanoacrylate adhesive. This has been particularly successful with fossils from the North Slope of Alaska. The low-viscosity cyanoacrylates can displace the water that often saturates bones and teeth that are contained in melting permafrost. We have been able to extract hundreds of specimens within an hour that would have been lost or would have taken days to collect using the adhesives that were available in the 1960s and '70s. Another important advance in fossil preparation under field conditions, in the Arctic environment, is the development of a range of new formulas of plaster that not only set up in cold wet conditions better than earlier formulas but also form much stronger jackets for fractured and delicate bones. The addition of epoxies and various fibers in the plaster mix greatly aided our work on the North Slope.

I have little doubt that consolidants with even greater properties of water displacement and adhesives with more strength and faster setup times will be developed in the near future. These and stronger quick-setting plasters will be a great boon to dinosaur fieldwork in the Arctic.

Permafrost Tunnels and Arctic Dinosaurs

Thom Rich and Patricia Vickers-Rich, the two colleagues from Melbourne, Australia, who landed at our base camp on the Colville River in 1989 to find out from me and my colleagues from Berkeley what had been learned since the last reports in 1987, happened to arrive during the warmest summer field season ever in my eighteen years of research on the North Slope. As temperatures reached the high eighties and lower nineties on the Fahrenheit scale for days on end, a chorus of loud crashing slides punctuated our days and raised our pulses as we excavated along the base of the Colville bluffs. When subjected to direct sunlight, the combination of melting permafrost and unconsolidated sediments at the top of the bluffs formed water-saturated slides that came noisily down the chutes and into the river with great splashes every fifteen to twenty-five minutes (see plate 4 for a vivid example). Permafrost can be found in rock, soils, and/or sediments. Permafrost can continuously underlie a whole region, as it does Alaska's North Slope, northern Siberia, and Arctic Canada. It can be found as scattered "islands" in Fairbanks, Alaska, and other parts of the subarctic interior of Alaska, Canada, and Siberia. Permafrost may reach depths of only a few tens of feet or 2,000 feet (610 meters).[4] Luckily, these slides mostly occurred when we were in our tents at base camp. When they did occur while we were mapping and excavating our quadrats, we would scramble or dive out of our individual "coyote holes" (see figure 4.2A) to avoid the rock and debris headed for us.

Once we experienced such an event, the possibility of another occurrence weighed on our minds and impacted our mapping and note taking. Thom and Pat came up with a possible solution. They had carried on excavations at Dinosaur Cove some 60 miles southwest of Melbourne on steep and very rugged ocean coastal cliffs. These challenging conditions forced them to literally use hard-rock mining techniques to free their dinosaur fossils by tunneling into the cliff face. This required air-powered jackhammers, mining drills, and special wedges to break the rock between drill holes.[5] The broken rock then had to be "mucked out" using shovels and hand tools. Eventually, the original entrance tunnel (adit) was expanded using explosives. Around a warm campfire, accompanied by some liquid spirits, Thom and Pat shared their experiences. This led the three of us to hatch a plan that might avoid the dangers of slides, escape the pesky mosquitoes, and prolong the field season—build a tunnel into the Colville bluffs just above the Liscomb Bone Bed.

Their Dinosaur Cove tunneling approach had been the first of its kind in the world of dinosaur digs; this Alaskan venture would be a first in the pursuit of dinosaurs in the Arctic. As we continued to brainstorm, we speculated that a tunnel might yield dinosaur bones in a better condition since the bones would not have been subjected to the yearly freeze-thaw cycle common in the surface quarry squares. Thom was especially keen about the increased safety that a tunnel would provide for us and our volunteers. Thom also emphasized that the tunnel would be dug into permafrost, which would probably not require an expensive shoring up as long as we had a proper temperature-resistant portal.

I was quite excited about the possibility of avoiding the restrictions raptor nesting during the summer put on our field seasons. If we were successful, the tunnels would allow us to excavate the bone bed from May to late September, depending on what kind of living shelters we would be allowed to station at the site. This would replace the narrow window of mid-July to the end of August set by snow cover and raptor nesting cycles. By the time Thom and Pat had departed the Colville, we had all resolved to find a way to garner support for this "revolutionary" approach to Arctic dinosaur excavation.[6]

This would ultimately prove to be much more difficult than I had imagined; however, I was buoyed up by Thom and Pat's track record in finding support for their "outside the box" approach at Dinosaur Cove. First I would have to convince my colleagues at the Bureau of Land Management that this approach was doable within their framework of regulations and that what we were proposing would lead to real scientific results, not just a tunnel in the permafrost. Having well-respected miners with permafrost tunneling experience on board our team would be helpful in persuading the Bureau of Land Management. We were able to identify several groups of reputable miners in Alaska that considered our plans to be well within permafrost mining "state of the art" in the 1990s. Thom and I connected with two mining partners in the Fairbanks area, Mike Roberts and Earl Voytilla, who not only had extensive experience with permafrost tunneling but also had worked on the North Slope with a boring machine known as a road header. They assured us that they could precisely bore into a bluff and keep within safe distances of our bone bed if the setting permitted. In addition, they took us to a permafrost tunnel near Fairbanks that they had dug to recover placer gold.[7] This gold mine had been completed over four years before we visited it and had remained stable with only an insulated portal and door.

After visiting the Roberts-Voytilla tunnel, I reacquainted myself with the U.S. Army's Cold Regions Research and Engineering Laboratory permafrost tunnel at Fox some 10 miles northeast of Fairbanks. This larger and longer tunnel had been dug between 1963 and 1969 by the U.S. Army and had served as a permafrost research facility. As part of the Arctic "cold war" between the United States and the Soviet Union, the tunnel had provided data for potential underground military sites in the Arctic, and its construction spawned the development of better engineering standards for roads and buildings constructed on permafrost in the interior of Alaska. In addition to being familiar with the Cold Regions Research and Engineering Laboratory tunnel, I had also visited a larger facility at Yakutsk in the Soviet Union. I was able to visit this facility as an exchange scientist under the auspices of the Soviet Academy of Sciences and the American Academy of Sciences during 1977. The Soviet Union had developed its facility in Yakutsk in eastern Siberia for many of the same military reasons as the United States had built its in Alaska. I was aware that both facilities required large refrigeration units to keep the tunnels at, or a couple of degrees below, the freezing point. Thom and I were delighted to learn that the Roberts-Voytilla tunnel did not require an expensive refrigeration unit to maintain the internal temperature, just a well-insulated portal. It became critical to our plan to occupy the tunnel for several field seasons and excavate fossils in order to acquire the latest data from the Cold Regions Research and Engineering Laboratory and the U.S. Bureau of Mines. We needed to know such things as the number of BTUs that individual miners typically expended per hour when working in a permafrost tunnel. It was also critical to find out what the heat expenditure rates for various types of equipment that we planned to use in tunneling and subsequent fossil excavation operations were. We were able to use the data provided by these agencies to come to the conclusion that we would not produce enough heat energy to degrade the permafrost needed to keep the tunnel from collapsing. Armed with this information, Thom and I were inspired to forge ahead with our plans for the Liscomb Bone Bed tunnel. All we needed now were the funds to make a more detailed feasibility study prior to launching the more challenging fund-raising campaign to actually dig the tunnel.

Our timing proved to be perfect. With the encouragement of Philip Currie, one of the founders of the newly established Dinosaur Society, Thom and I submitted a proposal to fund the first phase of our tunnel project—establishing the engineering feasibility of our plan. We were rewarded with one of the Dinosaur Society's earliest and largest grants of $10,000. This seed money provided the basis for a field consultation between Thom, me and the Roberts-Voytilla team in July 1994 on the Colville River and Liscomb Bone Bed. Within less than a day on the site, the Roberts-Voytilla team concluded that the "dinosaur tunnel" was feasible and well within their capabilities, as long as we could haul the road header and supporting equipment to the site along the fully frozen Colville River. This would mean that the work would have to be done at the end of the winter or beginning of spring, sometime around March or April. Delighted, Thom and I returned to Fairbanks to map out our strategy for phase 2—full funding for the tunnel construction. The cost of this venture, a minimum of $250,000, was much higher than most of my standard excavations. A typical season would cost $10–15,000 plus the time donated by volunteers and helicopter or other support given by private and federal agencies. Paleontological fieldwork in the Arctic is supported by relatively small budgets compared to most other areas of the geosciences and physics. We estimated that the project would cost between $250,000 and $500,000 in 1996, the year we planned to start the tunneling.

Raising this amount proved to be a daunting thirteen-year task. We approached ARCO Alaska, the National Science Foundation and its Australian equivalent, the state of Alaska, the Bureau of Land Management, the National Geographic Society, and even a group of investors led by former Alaskan governor Wally Hickel. If we hadn't been given so much encouragement by Philip Currie and several individuals in Alaska's mining industry, plus that extended by Robert King and Mike Kunz at the Bureau of Land Management, we would have given up long before 2007 when the "dinosaur tunnel" finally became a reality. In addition, we owed a great deal of thanks to William Hopkins, the executive director of the Alaska Oil and Gas Association, for his constant encouragement and provision of office space and access to important contacts in the oil industry. Thom and I received enthusiastic receptions at public lectures that we gave and at a number of presentations before mining and oil industry groups. However, no monetary support came forth. We also found ourselves being turned down consistently by all of the agencies and organizations that usually support scientific research in the United States and Australia. What we were up against was that the representatives of the larger agencies, such as the National Science Foundation, could not seem to get beyond the "risk" factor that they ascribed to working in a tunnel. Despite our enthusiastic presentations and supportive data, they couldn't seem to grasp the larger picture. Once we proved the method, it would allow for the excavation of numerous sites that were out of reach because of their precarious positions along the Colville bluffs. These could be reached by first sinking a vertical shaft and then extending tunnels from same. But first we needed to demonstrate the safety and effectiveness of the tunneling approach. We also had to show that proper mapping and excavation could be safely carried out in our first tunnel to prove the scientific merit of the approach. In 2003, I retired from my position at the university and museum in Fairbanks and withdrew from the floundering project. Thom vowed to carry on, and that he did with the tenacity and conviction that has been a hallmark of his career.

In March and April 2007, the combined efforts of some 144 individuals, eight companies and institutions, and more than $800,000 bore fruit: the long sought-after permafrost tunnel was finally dug.[8] In the end, it was not so much its scientific merit as much as its possible potential for attracting the public's attention as a documentary on dinosaurs that delivered the support needed to dig the tunnel. Most of the money was provided by a media combine put together by producer Ruth Berry. This media group consisted of ARTE France, NOVA USA, Screen Australia, and the Australian Broadcasting Company. The "dinosaur" tunnel was, in the end, dug using drilling, blasting, and mucking techniques. These were all standard hard-rock mining practices, and they were carried out by Robert Fithian and his crew rather than Roberts and Voytilla using their road

Fig. 8.1. Oblique aerial view of Colville River bluffs at high water in early June, tunnel portal near corner of lower right-hand quarter of photo *Credit: Kevin May.*

header machine. Fithian was another one of the recommended miners that we had been put into contact with during the planning phase of the project. The 10-foot × 10-foot tunnel reached some 30 feet into the bluff, positioned to keep the targeted bone bed as a floor. An added bonus was revealed in the tunnel's ceiling when a series of dinosaur footprints were exposed following the initial excavation. Unfortunately, these natural casts were lost when the surface of the tunnel ceiling swelled and exfoliated before photos could be taken. Because of the almost horizontal orientation of the rocks, footprints and trackways are hard to discern when standing at the outcrop, but within the tunnel, the footprint casts could be seen in 3-D relief. Prior to this discovery, my assistant Kevin May and I were convinced that some of the bedding contacts held tracks. Subsequent to the exposure of these dinosaur tracks, Paul McCarthy and his students recorded several trampled bedding surfaces as a result of a detailed stratigraphic and sedimentological study of the outcrops that included the Liscomb Bone Bed.[9] Excavation of the tunnel floor and bone bed had to await summer. A combined team of Australians and members of the University of Alaska Museum, including Kevin May, arrived in early August to find the floor of the tunnel buried in solid ice; high water had seeped in through the portal and then froze during breakup of the river ice. This possibility had been anticipated by Thom and had worried Mike Kunz of Bureau of Land Management. But with the aid of a jackhammer and compressor that Thom had brought along, the tunnel floor was laboriously cleared allowing the first excavation of fossils. The results, according to Kevin, were encouraging; the bones were generally in better condition than those encountered in earlier external excavations. Unfortunately, a rift developed between the Australian team and the Alaskan team regarding excavation techniques and disposition of some of the fossils. Mike Kunz and the Bureau of Land Management came down on the side of the Alaskan crew and the University of Alaska Museum. Thom withdrew to Australia a bit disappointed but satisfied that he had basically accomplished what he primarily came to do—establish the feasibility of the method and develop a novel approach to dinosaur digging, just as he had done in Dinosaur Cove, Australia. Kevin May and Patrick Drukenmiller of the University of Alaska Museum of the North have continued the excavations inside the tunnel and have recovered more bones and teeth. Future work in the tunnel is in doubt due to annual flooding and the lack of significant results. When all is said and done, I admired Thom's vision and tenacity. I appreciated the potential of the tunneling method to make many fossil sites in Arctic Alaska more accessible. However, the cost per excavation using the presently available techniques would keep the tunneling technique out of reach in the foreseeable future.

Tunneling to excavate dinosaurs does have potential for Arctic areas outside of Alaska. Dinosaurs in the Svalbard Archipelago and on Axel Heiberg are in topographic contexts that could benefit from the tunneling techniques that Thom and Pat developed in Australia and Alaska. Cost, again, would be a limiting factor, but changes in the economic picture could make this approach more feasible.

DNA and Permafrost: An Arctic "Jurassic Park"?

Could the process that produces permanently frozen ground (permafrost) also preserve significantly long portions of the DNA molecule or other biomolecules for millions of years, and even more to the point, over the sixty-six million years since dinosaurs walked the Arctic? Would the Liscomb Bone Bed permafrost tunnel be a path to such a discovery? Any discussion that combines DNA and dinosaurs usually conjures up the themes and images of the 1995 blockbuster movie *Jurassic Park*—never mind that the movie focuses primarily on dinosaurs from the Cretaceous and that it exhibits the typical exaggerations and misconceptions about dinosaurs and those who study them that Hollywood can't wean itself from. The movie and novel had profound effects on the general public's view of dinosaurs and what their study might lead to. Michael Crichton has to be credited with bringing the role of sciences such as molecular biochemistry, genetics, and animal ethology (animal behavior) in the study of dinosaurs and paleontology to the public's attention. In Crichton's story, dinosaur DNA is extracted from ingested dinosaur blood contained in mosquitoes and other blood-sucking insects preserved in amber. This was a quite an outlandish idea in the late 1980s, but if you look at more recent research on ancient DNA and other biomolecules you will see that what Crichton envisioned is now much less fantasy and much more possibility.[10] At its core, the movie pointed to the real possibility that 80 to 110-million-year-old fossils could contain ancient biomolecules that had been a part of the organism during

Fig. 8.2. Fithian mining crew drilling at entrance to the Liscomb Bone Bed tunnel in late March 2007. *Credit: Kevin May.*

its life. Before I turn to the Arctic of Alaska and permafrost, I want to offer a quick review of the research that led to the discovery of ancient DNA and other biomolecules.

For a large segment of the populace, this story was their first introduction to the general field of chemical fossils; however, this area of research really began in 1953 with the team of Nobel Laureates James Watson and Francis Crick. By deciphering the structure of the DNA molecule, they laid the foundation for the fossil biomolecular research that was to follow. In the 1960s, Melvin Calvin, a biochemist and Nobel Laureate, brilliantly elucidated the structure and function of chloroplasts in plants and went on to demonstrate that rocks millions and billions of years old could retain stable chemical remnants and markers for fundamental biomolecules such as chlorophyll. In the 1980s, Allan Wilson, a colleague of Calvin's in the University of California's Biochemistry Department at Berkeley was pursuing ancient DNA—that is, DNA derived from fossil microbes, animals, and plants that are tens of thousands to millions of years old. Wilson sought to elucidate the evolutionary relationships of extinct organisms by studying their DNA. This long and complex biomolecule contains a genetic "family

Fig. 8.3. Anne Pasch and Amanda Hanson excavating a quadrat inside the Liscomb Bone Bed tunnel in August 2007. *Credit: Kevin May.*

portrait" that points to the network of ancestors that preceded the individual whose cells now contain the DNA.Unfortunately, the DNA molecule is relatively fragile and unstable. DNA can be found in every living cell of all living organisms, from bacteria to humans. While the organism is alive, there are internal mechanisms that can repair damage and mutation, maintaining the integrity of the molecule. Upon death, these mechanisms are no longer functional and the molecule starts to degrade immediately. The environment within which DNA resides can, moreover, speed up or induce degradation. The presence of some bacterial enzymes can "digest," that is, break bonds within the molecule. The presence of free water with certain free ions speeds up breakdown. Acidic solutions (low pH) and temperatures above 99°F (37°C) also work to disassemble DNA.[11] However, there are several factors that aid in the preservation of DNA. Rapid burial or saturation with tar or resin, freezing temperatures, or an alkaline solution (high pH) all promote the preservation of DNA in ancient sediments. Therefore, the main challenge of finding ancient DNA is finding fossil organisms that die in very special environments and have undergone very restrictive histories of fossilization. Rapid burial in an oxygen-free, dry, and cold environment are more likely to result in the preservation of larger segments of the original DNA.

Since the chances of finding complete fossil DNA are minimal, then anything that could increase the abundance of disconnected pieces of the molecule would greatly aid in the hunt for these segments. Such a breakthrough came in the form of the discovery of polymerase chain reaction—a technique to amplify very small segments of the DNA molecule in order to help in mapping the details of the molecular structure that is present. Polymerase chain reaction was developed in the mid-1980s by Kary Mullins, an industry-based biochemist, who was awarded the Nobel Prize for chemistry in 1993. At this time, the San Francisco Bay Area was becoming a hotbed of ancient DNA research, as a husband and wife team, George and Roberta Poinar, joined Wilson at the University of California at Berkeley. The pioneering efforts of the Poinars and Wilson initiated widespread acceptance of the idea that at least some parts of DNA and other biomolecules could survive the vagaries of geological processes and be identified in fossils millions of years old.

By the late 1980s and early 1990s, I was steadily establishing my research program on the dinosaur record of northern Alaska, both in the field and the lab. During this time, it was discovered that fragments of ancient DNA could survive thousands and millions of years of burial in the Earth's sediments. Research on the stability and preservation potential of the DNA molecule opened up the possibility that the polar regions and permafrost might have imprisoned some fairly complete ancient DNA molecules. This possibility led to the discussion of the potential for cloning extinct Ice Age animals, such as the mammoth, using ancient DNA.[12] It also led me, in 1994, to cross the path of a graduate student at Montana State University, Mary Schweitzer. Mary was working on fossil biomolecules and was interested in obtaining dinosaur bones that had been encased in permafrost sediments. Two papers published earlier in the journal *Science*, had alerted her to attempts to extract DNA from Cretaceous dinosaur bone fragments.[13] Mary felt that permafrost-encased bone might produce better preservation of a range of biomolecules, including longer DNA strands. Mary, aware of my work on the North Slope, contacted me by email and asked if I could take some samples of dinosaur bone on my next field expedition to the North Slope. This proved to be difficult because digging through frozen sediments is a lot like digging through concrete. Usually, extracting fossils from the bone bed is done after the sediments have thawed. When thawed, the bone bed rock becomes less cohesive and is easier to split. Taking the samples required specific collection protocols to avoid contamination. There are several common sources of contamination in the collection process. To avoid contamination of stray skin cells from the collector or the introduction of bacteria and airborne fungal spores, I used disposable latex gloves, wore a mask and head covering, and then kept the samples frozen with "dry" ice, at which point they were shipped to Montana. To date, the biomolecular work on these materials has not been completed, but Mary and her colleagues have made some exciting progress with other biomolecules.[14] Perhaps the Liscomb Bone Bed permafrost tunnel will offer new opportunities to extract DNA from frozen dinosaur bone, as well as provide an even wider variety of ancient biomolecules than has been found thus far.

Before leaving the discussion of permafrost and DNA, I should clarify that most permafrost in the Arctic was formed over the last few hundreds of thousands of years and not during the Cretaceous, the time of the dinosaurs. However, the complete age range of permafrost in the Arctic is far from being fully deciphered. There may be some undiscovered permafrost deposits in the Arctic that will far exceed our present expectations and techniques. Greenland and other parts of the high Arctic such as Ellesmere Island appear to have the highest potential due to their high latitude and extensive ice caps. Perhaps a future graduate student will take up the challenge and produce significant results.

9 NATURAL RESOURCES, CLIMATE CHANGE, AND ARCTIC DINOSAURS

Arctic Dinosaurs and the Energy Crisis

What could Arctic dinosaur research and short-term solutions to our country's energy crisis have to do with one another? Can one burn dinosaur bones to produce energy? Are dinosaur fossils an important source of North America's petroleum? Are dinosaur bones an alternative source of energy to coal and petroleum? The answer to the first suggestion—burning dinosaur bones to produce energy—is "No!" The answer to the second (despite Sinclair's Oil's iconic trademark) is also a "No!" Interestingly, the answer to the third question is a qualified "Yes!"

During the great uranium "rush" of the 1950s in the Four Corners region of the southwestern United States, researchers found that Jurassic-age dinosaur skeletons often contained high amounts of uranium salts.[1] These concentrations of uranium were great enough to make the collected bones so "hot" that curators of museum collections were required to place them in lead-shielded containers or have them removed and sent to regional repositories for nuclear waste. This was actually one of my first chores as a curator of Earth Science at the University of Alaska Museum—an aspect of collection research and safety requirements with which I was totally unfamiliar prior to assuming this position. There is little doubt in my mind that some uranium-rich dinosaur bone ended up in reactor fuel rods or atomic bombs.

Arctic Dinosaurs and Coal

The real connection between the energy crisis and dinosaurs is related to coal. Coal will continue to be an important source of energy for at least the next decade. At present, around 50 percent of the electricity generated in the United States originates in coal-fired plants. Since the United States possesses between 25 to 30 percent of the world's coal reserves, and Alaska harbors 30 to 50 percent of the U.S. reserves, Alaska will become increasingly important as a source for this fuel in the foreseeable future.[2] A look at the hydrocarbon map (plate 8) reveals the widespread coal-mining sites in the circumarctic. Thus far, the most significant deposits of coal are found in Alaska. Most of Alaska's coal is mined near Fairbanks and is burned locally to generate electricity or is shipped to Korea. This may very well change as industrial and population growth in the United States and the Pacific Basin continues, especially in light of the fact the low-sulfur content of the

Alaskan coals meets some environmental standards that those mined in the lower forty-eight states can't meet. Another property of Alaska's coals that will make them attractive to future energy planning is their great potential for storage of methane or natural gas.[3] Methane produced from coal can be transported by pipeline over long distances with less technical difficulty than the solid coal. Presently, very little of Alaska's coals are being mined due to their remoteness.

Where are the richest dinosaur-bearing sedimentary rocks to be found in Alaska? The answer is in Early to Late Cretaceous coal-rich sedimentary rock units called the Nanushuk and Prince Creek formations (see figure 3.1). These formations are included within what the U. S. Geological Survey calls the northern Alaska-Slope coal province. This province is estimated to contain at least 80 percent of Alaska's proven coal reserves. This coal province also contains the vast majority of the dinosaur remains that have been found in Alaska. Dinosaur fossils and coal deposits go hand in hand.[4]

When you combine the coal, coal-methane, and petroleum reserves of northern Alaska, it is clear that the most prolific dinosaur-bearing areas of Alaska will be targets for extensive energy planning and exploitation in the next few decades. There will definitely be a significant impact on dinosaur research if the coal, methane, and petroleum resources of Alaska's North Slope become the focus of exploitation as part of a national energy policy. The good news is that it would most probably provide the monetary and technical basis for much more extensive paleontological surveys and excavation through grants from both state and federal management agencies. In addition, energy companies would probably be required to fund studies to evaluate the potentials for damage to paleontological resources in order to obtain permits. Agencies like the U.S. Geological Survey and the Alaska Division of Geological and Geophysical Surveys would be in a better position to fund the data gathering and assessment of the detailed geology and paleontology of the North Slope. Both of these agencies are presently engaged in cooperative field programs to produce the detailed geological maps and analyses that are needed to provide a sound basis for resource development. National and state energy companies would also want to expand their geological and paleontological databases, but much of this data would be proprietary and therefore not initially available to the public or noncompany-contracted paleontological researchers. However, past history suggests much of the proprietary data would eventually become available.

The increased resources that would be available with increased commercial interest would likely permit researchers to get to dinosaur fossils that are presently inaccessible. Expensive techniques such as permafrost tunneling that was applied to the Liscomb Bone Bed would allow researchers to excavate dinosaur-bearing beds exposed in the middle of high or precipitous bluffs. Permafrost-supported tunnels, unlike the Liscomb Bone Bed tunnel, could be accessed by vertical shafts that would reach the dinosaur-bearing levels. Therefore, the potential positive results of increased focus on the North Slope carbon-based resources would be field research initiatives that could provide access to parts of the rock sequences that are now impossible to reach safely.

The negative side of the ledger is that easier access and the increased information that would attend resource development could draw amateur and commercial fossil collectors to the North Slope. The Bureau of Land Management has already had to deal with such collectors under the present difficult field conditions. The increased presence of construction and maintenance crews during the summer could add to untrained and unauthorized collection of dinosaurs as well as of Pleistocene fossils such as mammoths. These negatives could be controlled by vigorous and well-trained regulators from state and federal management agencies charged with oversight and equipped with relevant and effective laws and regulations. Presently, there are very few properly trained paleontologists on staff at agencies such as Alaska's Department of Natural Resources or federal agencies like the Bureau of Land Management. This should change over the next few decades as the Arctic becomes the site of the new "cold war" that will pit circumarctic governments against one another as energy and other resources become even scarcer.

Coal, Amber, and Ancient DNA

Amber, the gem-quality form of fossil plant resin, is found in a range of sedimentary rocks from sandstones to lignites that represent marine to alluvial environments. A lower-grade form of fossil resin called retinite is often concentrated in lignite. Arctic Alaska has extensive deposits of lignite, a low-grade form of coal, that are rich in retinite and amber. These lignite beds are often found interbedded with dinosaur bone beds and other alluvial deposits that are dinosaur bearing.

This relationship between coal and dinosaurs can be brought to bear on the importance of amber in the search for fossil DNA. The mid-1990s witnessed a new direction for paleontology and ancient DNA research—the extraction of DNA from organisms enclosed in amber. In particular, fossil insects found in amber were proving to be a very promising source of ancient DNA, and as Michael Crichton imagines in his cautionary tale *Jurassic Park*, blood-sucking insects might even harbor the DNA of dinosaurs who acted as blood suppliers (see plate 9).[5] This new focus brought together biochemists, entomologists, and paleontologists specializing in invertebrates. California remained a major focal point with the teaming up of Raúl Cano at the California Polytechnic State University at San Luis Obispo and George and Roberta Poinar in Berkeley and their son Hendrik, a graduate student of Cano's. Paleo-entomologists such as David Grimaldi at Cornell University were actively studying the rich insect life trapped in amber that was aided by the concomitant preservation of soft tissues and internal organs. Fossil insects and other organisms are first trapped in tree resin and their bodies are rapidly sequestered from the atmosphere and degradation by acid waters and oxygen. The resin contains chemicals that kill bacteria, therefore shutting off their enzyme production. The resin quickly polymerizes into a protoamber material called copal. Polymerization hardens the resin and helps to further lock out oxygen, water, and microorganisms. Further polymerization after burial in sediments then turns the copal into amber.[6] All of these attributes of the amber-forming process work to preserve the original DNA that is part of the entrapped organism.

The study of amber has touched my career at two critical times. In the spring semester of 1961, I was a new and eager graduate student in paleontology at Berkeley. I was fortunate to be invited by Wyatt Durham to participate in an expedition to Chiapas whose primary purpose was to study the geologic context of the extensive amber deposits. The Chiapas amber was just beginning to be recognized for its great abundance and diversity of enclosed fossil organisms.

That summer in Chiapas was a great learning experience for me as a student. Ultimately, it also played a role near the end of my career when I was deeply involved in Cretaceous dinosaur research at the University of Alaska Museum in Fairbanks. The museum collections that I was responsible for contained, I discovered, extensive amber specimens from all over the world. Most importantly, I found that there were specimens of amber from many parts of Alaska and that most had been collected in northern Alaska from Cretaceous-age formations. The vast majority of the known amber deposits are either Mesozoic (Cretaceous) or Cenozoic (Paleogene) in age. Most of the Alaska amber is found in the mid- to Late Cretaceous-age rocks that are widely distributed over the western third of the North Slope (see figure 3.1). I now had a good reason to renew my interest in learning more about amber.

In 1960, a year before Robert Liscomb collected the first dinosaur remains in Alaska, Ralph Langenheim, a member of the 1961 Chiapas expedition, and his colleagues discovered and reported on Cretaceous-age ambers from deposits along the Kuk, Kaoluk, and Ketik rivers of the Arctic Coastal Plain of northwestern Alaska.[7] Although the presence of amber and other fossil resins in Alaska had been known for some time, this was the first scientific documentation of the presence of insect-bearing amber from Alaska.

Some forty years later, Langenheim's report on Cretaceous insect-bearing ambers in Alaska caught the attention of David Grimaldi, now at the American Museum of Natural History in New York. David contacted me and Donald Triplehorn, a Fairbanks colleague and noted expert on coal deposits and closely related materials such as retinites. Grimaldi was primarily interested in the insect-bearing aspect of these Arctic ambers. Langenheim and his colleagues had established the presence of four insect families in the Alaskan resinites. Subsequent research by several state and federal geologists had reported the presence of dinosaur fossils in the rock formations that contained coal and the insect-bearing ambers.[8] Grimaldi was most interested in the evolution of the insects that were found in the amber. The co-occurrence of amber and Cretaceous dinosaurs in the same coal sequences sweetened the "pot" for me. Grimaldi and his team from the American Museum of Natural History didn't find any insect-bearing amber on their first try on the Kuk River. However, they were encouraged by the abundance and size of the amber pieces that they came across. There are a lot of Mesozoic-age amber-bearing deposits in Alaska that Grimaldi didn't have the time and money to investigate. These undiscovered treasures may very well contain a rich record of insects trapped in fossilized tree goop, and some of the sequestered insects may very well contain pieces of dinosaur DNA. Cretaceous amber found in southern Alberta has yielded a variety of insects, including a mosquito.[9]

This chapter of Alaska's paleontological story is far from closed. Hopefully, other researchers will eventually follow Triplehorn and Grimaldi's pioneering pathway and will bring forth exciting and highly significant results to the world of fossil biomolecules. Will we ever be able to clone and raise a dinosaur from the dead? The Poinars, Grimaldi, and others who pursue ancient DNA do not adhere to the common definition of biologic death. They have a more expansive and intriguing definition that is based in biological evolution. If one takes it that DNA is the fundamental genetic molecule of life and that it carries pieces of all of the life that preceded it in time, then there is no death until all the DNA that exists is totally disassembled and scattered throughout the cosmos. A simpler and more common recognition of this can be seen in the expression that one often hears at funerals—that person is not really dead but "lives on in their children."[10]

Petroleum in the Arctic

Petroleum resources may not be as intimately tied to dinosaurs studies as coal is, but when it comes to the world's feverish quest for energy sources, it certainly sits as an equal on the resource stage.[11] At present, the circumarctic has four important petroleum production and/or potential production areas. From Alaska around to the west these include northern Alaska and northwest Canada, the Lena Delta and Laptev Sea shelf province, the Barents Sea province, and the Sverdrup Basin province (see plate 8).[12]

Of the four important petroleum areas in the Arctic, northern Alaska, the Barents Sea, and the Sverdrup Basin have the greatest near-future potential for impacting Arctic paleontology in general and dinosaur studies in particular. All of these areas have produced dinosaur fossils. Northern Alaska has the greatest potential for the positive and negative impacts on dinosaur research with respect to petroleum exploration. This is primarily because of the relative ease of access compared to the Sverdrup Basin and Barents Sea areas. However, as the Arctic continues to warm and the sea ice, icecaps, and glaciers retreat, the sea lanes will expand, and access to the complex of Canada's Arctic islands and their interiors will increase (see figure 2.1 and plate 8). This will greatly facilitate prospecting for petroleum and dinosaurs. It will, like other precious resources, open the Arctic seafloor to territorial disputes over ownership. Circumarctic governments and private companies will pour resources into the region. A positive outcome could be the compilation of more complete geologic maps and ground data. More detailed data in the Svalbard-Barents Sea shelf and Greenland subregion could provide a sound testing of the Barentsia (see figure 6.1) intercontinental dinosaur migratory pathway hypothesis and greatly expand our knowledge of Cretaceous dinosaurs and the paleoecology of the eastern Arctic.[13]

So, the bottom line is that there is a very good chance that Arctic Alaska, as well as the rest of the circumarctic, will be undergoing significant changes because of energy resource exploitation over the next few decades and that these changes will have a mix of positive and negative effects on the Arctic dinosaur record. The increased quest for Arctic carbon-based treasure fired by the looming world energy crisis is not the only process that will have a significant impact on the Arctic in the near future. Global climate change is already here and the Arctic is the metaphorical "canary in the coal mine" that has alerted us to the problem.

Global Climate Change Renews Interest in the Arctic

Even though dinosaur research in the Arctic over the last twenty-three years has produced an impressive variety of taxa and now boasts of an extraordinary quantity of specimens, this only represents the tip of the iceberg (see figure 2.1). The limitations and demands of the Arctic environment mean that paleontological progress proceeds at a slow and costly pace. I hope that I have clearly communicated the physical and mental challenges that fieldwork in the Arctic presents to the paleontological researcher. What stands out from my experience in Alaska's Arctic is a combination of factors. There are the short field seasons, the relative lack of rock exposures, the rapidly changing weather, and the great unpopulated distances that must be traversed. A field paleontologist is pretty much confined to using navigable rivers and streams, of which there are surprisingly few, even if you use inflatable shallow-draft craft. Overland travel on Arctic tundra is exceedingly difficult and slow in the summer due to the density and character of the tundra vegetation and the numerous thaw lakes formed by melting permafrost. Thaw-induced polygon nets greatly impede walking across the tundra and the alluvial plains (see figure 3.2 and plate 5). Since boat and overland travel are so limited, many of the rare and critical rock exposures must be prospected and studied with the aid of a helicopter. This is a terribly expensive proposition, and helicopters have short ranges and are notoriously prone to breakdowns. They require caches of fuel throughout any large remote area such as most of the North Slope. I don't know of any region outside of the Arctic and Antarctic that combines all of these limiting factors and challenges. It now appears that field research in the Arctic has reached a critical point in regard to the challenges that this region has presented to field researchers over the years.

The Arctic is being impacted by global weather change that appears to be the result of overdependence on petroleum and coal-based energy sources primarily in the Northern Hemisphere. Science over the last fifty years has established that the Arctic is warming and that sea ice and permafrost are both slowly disappearing. As the Arctic continues to warm, travel throughout the Arctic's waterways will be greatly facilitated by the retreat of sea ice and glaciers, but overland travel will most probably be made more difficult as the vast areas of permafrost degrade and the resultant marshes, mires, and thaw lakes increase in abundance and size. How long this trend will continue is anybody's guess at this point. But to paraphrase a popular saying: it is a foul wind that blows no one well.

Climate change will certainly stimulate important advances in alternative energy development, but make no mistake, the fossil fuels will continue to supply the bulk of the world's electrical and vehicle-generated energy for at least the foreseeable future. Therefore, human-induced factors that contribute to climate change will also persist into the foreseeable future. The vast coal, oil, and natural gas reserves of the circumarctic region (see plate 8) will continue to be the target of the international oil companies that have directed the course of energy use on this planet since World War II. This will have a profound effect on the future alignment of geopolitical blocs, on scientific cooperation, and on the potential for open conflict over the Arctic's resources.[14]

Where Do Dinosaurs Fit In?

What will this renewed interest in the Arctic mean to field geology and paleontology? As the United States and other countries ease back on their south polar commitments and refocus on the Arctic, much more money, logistic support, and interest in Arctic science will result. This is especially true for the U.S. part of the Arctic because of the disproportionate support that the United States has directed to the Antarctic over the last forty-five years and the recent realization that the Arctic waters and vegetation are important carbon-sequestering and pollution sinks for the highly industrialized United States and southern Canada. Even though there is potential for conflict between the nations that divide the circumarctic politically, there are also positive signs that these nations are beginning to recognize the Arctic environment's critical role in the Earth's ecosystem. Carl Benson, a glaciologist and a senior scientist in residence at the Geophysical Institute in Fairbanks, Alaska, notes that "ignorance of the Arctic is an infinite resource."[15] It seems that the general lack of critical knowledge of the Arctic has already motivated the Arctic-based nations to establish a wide range of organizations and cooperative environmental monitoring programs, namely, the Arctic Council, the International Arctic Environmental Data Directory, and Arctic Climate Impact Assessment. A recent positive effort by scientists to better understand the Arctic environment and its vulnerability to global climate change is reflected in the recent completion of the Third International Polar Year in 2009. This international collaborative effort was a nongovernmental, broad-based research program led by international scientific bodies including the Conservation of Arctic Flora and Fauna and the American Polar Society.

The geosciences have the tools and knowledge to compile the primary "deep time" record of the physical and biological evolution on this planet. Paleontology can no longer be portrayed as "stamp collecting," as some of physical science practitioners try to portray it. The integration of molecular biology, geobiochemistry, geochronology, and evolutionary biology, along with geophysics, has transformed the way paleontologists work and view the Earth's history and the way research results are applied to present workings of society and technology. If the warming of the Arctic continues for at least five more decades, paleoecologists and paleogeographers will have the opportunity to study the collapse and reordering of a major ecosystem with attendant extinctions, migrations, and reconstructions of food webs. This, along with research on coral reefs and tropical rainforests, will present an unprecedented set of case studies within a human lifetime. Paleontologists may finally get a firsthand view of what triggers major extinction events. Hopefully we, along with neontological biologists, will survive to tell the tale.

10 FUTURE EXPANSION OF THE ARCTIC DINOSAUR RECORD

The Colville River: The Red Deer River of the Arctic?

The southern Alberta buffalo plains greet you with their vast grain and forage fields, slight topographic undulations, endless skies, scattered ranches, and small sleepy towns as you proceed eastward from the hustle and bustle of urban Calgary. If you had no previous knowledge of the region's geography, within an hour you would find yourself trying to fend off the boredom of what seems to be endless flatlands that characterize most of the 90 miles (145 kilometers) to Drumheller. When you finally see the sign that directs you towards Drumheller, you turn north and slowly descend through a series of roadcuts that fail to stimulate even the ardent field geologist. However, this soon changes in dramatic fashion as you reach the outskirts of the small town of Drumheller and the gently meandering Red Deer River. The stacks of sedimentary strata interspersed with dark lenses of coal, lens-shaped ancient channel sands and conglomerates complexly sculpted into labyrinthine badlands delight even the jaded geologist's eye. Drumheller is about midway along the Red Deer River, which winds its way east, then south, then east again for over 400 miles (650 kilometers) as it seeks a confluence with the Saskatchewan River. This incised river valley was host to important early twentieth-century coal mining operations. It is now the heart and soul of Alberta's Cretaceous dinosaur country. This is where the magnificent Royal Tyrrell Museum of Palaeontology is to be found nestled within the Red Deer River badlands, just a few miles to the northwest of the center of Drumheller. If you follow the Red Deer River as it winds its way over 100 miles (163 kilometers) southward from Drumheller, you will come upon Dinosaur Provincial Park.[1] The park with its dinosaur research station, labs, and outdoor dinosaur exhibits is, like Drumheller, set within the spectacular Red Deer River badlands. Whether you are a paleontologist or a "dino" tourist, the Dinosaur Provincial Park will exceed your greatest expectations and impress you with its extraordinarily rich record of dinosaurs and the world they roamed in. This is why the Park was designated by UNESCO as a World Heritage Site.

Fieldwork between Drumheller and the Dinosaur Provincial Park over the last hundred plus years has amassed a record of the Late Cretaceous world and its dinosaurian inhabitants that is unmatched in the rest of North America. At least eighty dinosaur taxa have been identified so far.[2] Therefore, any time you find a dinosaur from the Cretaceous in western

Fig. 10.1. Aerial view the Royal Tyrrell Museum of Palaeontology and the surrounding Red Deer River badlands. *Credit: Courtesy of the Royal Tyrrell Museum, Drumheller, Alberta, Canada, modified by Roland Gangloff.*

North America, you must compare it to the record from Alberta early in your attempt to identify genera and species. It is also necessary to compare any new dinosaur faunal assemblages to the detailed biostratigraphic record that has been established for Alberta. The collections housed at the Royal Tyrrell and Dinosaur Provincial Park act as standards of excellence, and the institutions provide extensive collections of publications and the fine science that is contained therein. So in 1989, when my colleague Thom Rich exclaimed, "I think you may have the Red Deer River of the North here on the Colville!" he was referring to the richness of the dinosaur record from this part of Alberta rather than the physical setting. When Alaska's Colville River dinosaur assemblage and its geologic and environmental context are compared with the late nineteenth- and early twentieth-century history of the middle Red Deer River valley rather than with the present, it becomes clear that they have much in common.

During the early history of the Red Deer River collecting expeditions, sites had to be studied and collected from boats due to the remoteness from human settlements and the lack of roads and good trails. The lack of access roads is especially daunting since excavation of dinosaur bones usually requires the inclusion of surrounding rock matrix and the use of heavy plaster jackets to protect the fossils during removal and subsequent transport. Most prospecting for, and collection of, dinosaurs on the North Slope of Alaska is done with the aid of small boats. Even though helicopters and fixed-wing aircraft are available, fixed-wing aircraft have very limited landing sites, and helicopters are usually prohibitively expensive given the present level of funding. Both the Red Deer River and Colville River areas abound in remarkable concentrations of alluvial sediments and concentrations of dinosaur remains known as bone beds. Both areas share hadrosaur, ceratopsian, and theropod taxa, especially in Edmontonian time.[3] The hadrosaur *Edmontosaurus* is abundant in both areas, but a species assignment is still to be determined for specimens from the North Slope. The same is true of the ceratopsian *Pachyrhinosaurus* and the theropods Troodon, Albertosaurus, and other tyrannosaurids. *Dromaeosaurus albertensis* and *Saurornitholestes langstoni* appear to be conspecific faunal elements.[4] Thus far, the diversity and sheer number of dinosaur taxa that characterize the Red Deer River area are just hinted at on the Colville River. It must be remembered that in proportional area covered and person hours spent, the North Slope dinosaur research has barely just begun. However, the logistical costs due to the remoteness of the vast majority of Alaska's known and potential dinosaur-producing areas will always temper the pace of discovery and collection despite the possible counterinfluence of increased resource demand and exploitation.

Expanding the Arctic Dinosaur Record: Looking from East to West

Based on the most up-to-date geologic maps combined with the known records of dinosaurs, the part of the Arctic that has the greatest potential for significant future discoveries extends from northwestern Ellesmere Island in the Canadian Arctic Islands to almost the westernmost coast of northern Alaska. The Canadian part of this swath encompasses most of the Canadian Arctic Islands and the northernmost part of the mainland to the south—a vast area known as Nunavut. The portion of Nunavut that most likely will provide new discoveries of dinosaurs lies within the Sverdrup Basin. In particular, the northeastern part of the Sverdrup Basin, and more specifically the northeastern part of Ellesmere Island and adjacent Axel Heiberg Island, will most probably produce the most extensive and exciting dinosaur discoveries over the next two decades. These new finds will most likely coincide with increased oil and natural gas exploration throughout this region since the Sverdrup Basin exhibits the greatest potential for petroleum resources (see plate 8).[5]

The Canadian area directly to the west that touches the northeastern border of northern Alaska is part of the Northwest Territories and enfolds the natural gas-rich Mackenzie Delta. This western half of the Northwest Territories includes an area greater than that of the North Slope of Alaska. This part of the Canadian Arctic coast and contiguous area have not produced any dinosaurs yet. However, dinosaur remains were found farther south (chapter 2 and figure 2.1) in the District of Mackenzie on the East Little Bear River just south of the Great Bear Lake and the Arctic Circle.[6] The dinosaur remains were found in the fluvial-dominated Summit Creek Formation. This area is a promising coal-, oil-, and natural gas–producing area and is targeted for future petroleum and coal development.[7]

Preliminary geologic mapping and prospecting for fossils in the Peel River drainage has produced hadrosaurid and basal ornithopod remains in part of the Upper Cretaceous Bonnet Plume Formation. The Bonnet Plume Formation was formed in complex fluvial environments including floodplain, channel, and paludal conditions. Rock sequences have been identified as representing the correct paleoenvironmental conditions and therefore mark this area as having a high potential for further dinosaur discoveries.[8] Again, the high potential of this area to produce carbon-based fuels will expand exploration and geologic mapping efforts. Hopefully, this economic stimulus will bring in a greater number of bright, vigorous, and experienced field paleontologists to supplement the fine efforts of Grant Zazula and David Evans in the near future. This part of the Western Interior Seaway badly needs to be better constrained with fossil evidence.

The Northern Alaska Dinosaur Hunting Grounds

Northern Alaska, above the Brooks Range, encompasses some 80,000 square miles (over 200,000 square kilometers) of potential dinosaur-bearing rocks, spread over much of the southwestern and south-central parts of this region, with some outliers to the north and east (see figure 3.1).[9] Presently, only about 1,000 square miles (over 2,600 square kilometers) have been lightly prospected for dinosaur fossils, and less than 100 square miles (260 square kilometers) have been prospected or collected in detail. The area that extends from Ocean Point to Umiat in the north-central part of the North Slope remains the most thoroughly studied and collected part of northern Alaska and has produced over eight thousand specimens of dinosaur bones and teeth. Although this is a small fraction of what the Royal Tyrrell and Dinosaur Provincial Park museums hold in their collections, it far outstrips, by over an order of magnitude, all of the collections assembled in the polar regions in both hemispheres combined (see figure 2.1). Given the amount of land area contained in northern Alaska and the abundance of Cretaceous, dinosaur-friendly rocks, there is a mind-boggling potential for expanding the Cretaceous record of paleo-Arctic dinosaurs in Alaska. This area also provides great promise for adding to the skimpy record of other Cretaceous vertebrates such as mosasaurs, bony fish, and some types of birds that thrived in the Cretaceous Arctic seas, lakes, and estuaries. Only time, and much more work, will tell if Thom Rich's 1989 prognosis is to come true.

The western two-thirds of Alaska that lies north of the Brooks Range holds the greatest potential for increasing the Arctic record of Alaska. Aside from the rich record of dinosaurs that lies along the western bank of the lower Colville River, more work along the middle part of the Colville River has extraordinary potential for yielding new discoveries in rocks of the Upper Cretaceous to Paleogene Prince Creek Formation. The thicker and more geographically extensive Lower to Upper Cretaceous Nanushuk Formation (see figures 3.1 and 4.16) should be a prime target for future paleontological prospecting, specifically along the upper part of the Colville River, from the confluence with the Killik to that with the Awuna. This stretch of the Colville River and its tributaries, such as the Chandler and Killik, have extensive bedrock exposures and have already yielded enough dinosaur trackways to mark it as of prime potential for future discoveries of dinosaurs.

There are two localities that are found along the route of the Trans-Alaska Pipeline and Dalton Highway that are especially in need of careful and exhaustive prospecting. The first locality is Slope Mountain, which is of particular interest because it has extensive exposures of the Lower Cretaceous part of the Nanushuk Formation.[10] Slope Mountain is readily accessible from the Dalton Highway. The nonmarine deltaic and delta plain deposits exposed here are paleoenvironmental equivalents to the dinosaur trackway-rich Killik bend site discussed in chapter 4. The second is found along the western bank of the Sagavanirktok River, with exposures of rocks that represent the uppermost Cretaceous part of the Prince Creek Formation. The rocks along this part of the Sagavanirktok River form a series of steep exposures known as the Sagwon Bluffs. The much heralded K-T (K-P) boundary is most likely exposed in these bluffs, while it remains buried below the surface on the North Slope.[11] The Sagwon Bluffs were an intriguing target for research during most of my tenure in Alaska but were kept out of reach by raptor nesting restrictions. Both of these localities are logistically the easiest to reach and study since they are very close to the Dalton Highway and airstrips at Galbraith Lake and Happy Valley. There is also a University of Alaska research station close by at Toolik Lake. Hopefully, the Sagwon Bluffs will be less restricted by raptor nesting concerns in the near future.

The Alaskan Peninsula

An extension of the Alaska Range that eventually transforms into the Aleutian Islands thrusts itself out into the northern Pacific Ocean south and a bit west of Anchorage. If you venture out some 200 to 500 miles (320 to 800 kilometers) into this remote and mountainous peninsula, you will find a wilderness replete with active volcanoes and the largest concentration of brown bears to be found in North America.[12] You will also find thick sequences of Jurassic- and Cretaceous-age rocks that harbor a rich assortment of ammonites and other fossil marine mollusks. Until the late 1970s, no dinosaurs were reported from this area. With gas and oil finds in the Cook Inlet to the north, teams of exploration geologists began to scour this area, mapping and sampling the rocks as never before. This intense study quickly produced the first evidence (theropod tracks) of dinosaurs in the Late Jurassic Naknek Formation near Black Lake.[13] However, it wasn't until 2004 that this first dinosaur find was added to the North Slope collection (see figure 2.1).[14] Several footprints attributed to an indeterminate hadrosaur were found in the Aniakchak National Monument and Preserve. While working with the National Park Service at Katmai National Park and Preserve and closely related areas from 1999 to 2002, Tony Fiorillo and I came away convinced that this part of the Alaskan Peninsula will eventually become an important source of Jurassic and Cretaceous dinosaur remains. Subsequently, Tony turned his attention to the Denali National Park and Preserve, where he has added significantly to Alaska's dinosaur record.

Finally, recent restudies of two areas in northeastern Siberia have produced significant new dinosaur materials as well as more detailed contextual data. Both locales are below the Arctic Circle today but were most likely at or above the paleo-Arctic Circle when dinosaurs were living there.

The more exciting of the two is to be found along the Kakanaut River in the southeastern part of the Koryak Upland not far from Lake Pekulneyskoje in the Russian Far East (see chapter 2 and figure 2.1).[15] What was first appraised as a scrappy and impoverished accumulation of dinosaur fossil debris that was interbedded with thicker marine sediments has recently been shown to be a rich assemblage of floral remains accompanied by a much richer dinosaur fauna that includes dinosaur eggshell. The paleolatitude, the richness of the paleoflora (thirty taxa), and the significantly rich dinosaur assemblage in the latest Cretaceous rocks makes this locality particularly important in the reconstruction of the events that preceded the Cretaceous-Paleogene extinctions in the paleo-Arctic.[16] What other new and surprising finds in the Late Cretaceous rocks of the Koryak Upland might challenge our present understanding of this critical part of the Earth's history?

Hundreds of miles to the east beyond the great Verkhoyansk Mountain Range (see figure 2.1) is a locality on a small tributary of the Kemp'endyayi River in the Vilyuy Basin of northern Sakha (Yakutia). This paleo-Arctic locality, like that at Kakanaut, contains a greater wealth of fossils, including dinosaurs, than was first anticipated. It is unique among paleo-Arctic localities since it contains an assemblage of dinosaurs from the Late Jurassic and possibly the earliest Cretaceous.[17] Since its discovery in 1960, several expeditions have been mounted, the last in 2008. Each return has resulted in an increase in the volume of dinosaur material as well as an increase in the number of taxa represented. If the last forty-years years is any indication, the Kempendyayi-Teete locality surely holds a rich treasure of dinosaurs and associated materials that will greatly expand our knowledge of paleo-Arctic dinosaurs. In addition, this area will give important insights into the evolution of paleo-Arctic dinosaurs and, more specifically, their adaptations to the biologic challenges of these high latitudes.

Even though I have primarily stressed the physical challenges and limiting factors that field paleontologists face in the Arctic, I must end my discussion by stressing how rewarding and exhilarating field work in the Arctic can be. There is so much that remains to be discovered, recovered, and described from the dinosaur record in the Arctic. Only fifty years ago, the popular image of dinosaurs had them confined to the lower and middle latitudes, existing as amphibious dwellers in the rivers and lakes of tropical rainforests—slow, lumbering, tail-dragging beasts that were the icon of evolutionary failure. This erroneous caricature has been replaced by an exciting new view that is based on paleontological research that has documented the distribution of nonavian dinosaurs from the south polar region to very near the present North Pole.[18] No other land-dwelling wingless vertebrate can match this latitudinal distribution. Modern paleontological research has broken the shackles of confined thinking and reconstructed an ancient world that was populated by dinosaurs that evolved to be able to live and prosper in almost any terrestrial environment on Earth.

Work in the polar regions over the last fifty years has begun to fill in one the most important parts of the Earth's evolutionary history. It is in the polar regions, and especially the Arctic, that the next generation of paleontologists will find the fossils and contextual data that will fine-tune our compilation of the life and habits of these highly successful animals and finally supply answers to some of the more vexing questions concerning the great K-T (K-P) extinctions. If you want to make your mark in Mesozoic paleontology, go north, young men and women, for the field is wide open and there is so much to be discovered. Before you do, however, make sure you are up to the physical and mental challenges that this world presents.

Areas Outside of the Present Arctic That May Be Included in the Future

Over the period from 1925 to 2006, an impressive record of Late Cretaceous (Maastrichtian) dinosaur remains have been discovered in the Amur River Region of far eastern Russia between 48.5°N and 50.25°N.[19] These sites are just north of the border with People's Republic of China. More than fifteen hundred bones and teeth have been collected representing lambeosaurines, a nodosaurid, and theropods. The dinosaur fauna is dominated by a corythosaur-like lambeosaurines. More than 90 percent of the skeletal material is attributed to the taxon *Amurosaurus riabinini* and forms monodominant bone beds that are comprised of a mixture of mudstone, conglomerate, and diamicts. These bone beds are interpreted as terrestrial gravity flow deposits. Lesser concentrations of skeletal elements are found in coarse alluvial deposits.[20]

The most recent field studies (2005 and 2006) have been conducted just west of the city of Blagoveschensk on the northern bank of the Amur River. An international team of scientists from the Royal Belgian Institute of Natural Sciences and the Amur Complex Integrated Research Institute of the Far Eastern Branch of the Russian Academy of Sciences has made new excavations that yielded more than five hundred bones and has compiled a detailed taphonomic context for these.

The Amur dinosaur localities are important to the study of high-paleolatitude dinosaurs and their environment. These localities contain extensive faunal and floral evidence from high paleolatitudes some 900 miles southeast of the Kempendyay-Teete locality and were most probably within the subarctic latitudes some 10 degrees farther north than they are today during the Late Cretaceous (Santonian).[21] Palynological evidence and comparison to North American dinosaur faunas indicate that the Amur dinosaur faunas are Early to Middle Maastrichtian in age. Comparisons with coeval high-latitude dinosaur faunas in Alaska reveal intriguing distinctions and similarities. The hadrosaur faunas are dominated by late juveniles and few adults, but unlike the Alaskan faunas, lambeosaurines (crested) dominate rather than hadrosaurine hadrosaurs. Thus far, no unequivocal lambeosaurines have been documented in Arctic Alaska. Large-scale geological maps of this region between the Amur sites and the Kempendyay-Teete site indicate extensive outcrops of mid- to Late Cretaceous rocks. It is very likely that the area between the Amur Region and the Viluyuy River Basin will produce a significant number of Late Cretaceous dinosaur sites when resources become available to properly prospect this remote and tantalizing part of eastern and southeastern Siberia. Rocks older than the Mesozoic in this area have produced some oil but even more impressive natural gas plays.[22] Future geologic exploration in this entire region is a high priority with the Russian government and will likely bring geologic field parties and resources that are presently scarce.

GLOSSARY

Alluvium/alluvial: Pertaining to sediments deposited by rivers and streams.

Amber: A group of fossil resins derived from coniferous trees. Also known as resinite. Amber is usually light brown to light yellow and transparent to translucent and has a greater hardness, density, and higher melting point than the subfossil form copal. These physical properties are the end result of the polymerization of the molecule into long strands of C:H:O.

Angiosperm: Flowering plants, which is a large division of plants in which the seed is contained within a fruit. Includes plants such as citrus, palms, elms, oaks, and grasses.

Articulation: As applied to paleontology, it is the way skeletal components mechanically relate to one another. A fossil vertebrate skeleton is articulated when all of the bones are in their correct sequence and in contact with one another just as they were in life.

Biostratigraphy: The geologic specialty of dividing rock units and correlating them using their fossil contents. Biostratigraphers carefully trace the stratigraphic ranges of fossil taxa and use the ranges and assemblages to discriminate time units. Time units are based on the evolution of taxa and the principle that evolution of specific taxa cannot be repeated.

Bone bed: A rock layer that contains an unusually dense accumulation of bones, bone fragments, teeth, or other skeletal components representing more than one individual. Most bone beds have densities greater than twenty skeletal elements per square yard but can have greater than one hundred skeletal elements per square yard.

Brunton compass: A magnetic hand-held compass equipped with a mirror and folding site that can be corrected for magnetic declination named for its inventor, David W. Brunton,. Also known as a pocket transit. The Brunton also has a spirit level and inclinometer. You can always recognize a geologist in the field for they commonly have their Brunton "holstered" on their belt.

Calcite: An abundant, naturally occurring form of $CaCO_3$. This mineral commonly acts as a cement that binds sediments and helps to form sedimentary rocks.

Carnosaurs: The Carnosauria are stem-based taxon; they were theropods that were closer to *Allosaurus* than to birds. Includes taxa such as *Allosaurus, Cryolophosaurus, Carcharadontosaurus Acrocanthosaurus*, and *Sinraptor.*

Cenozoic: A major division (era) of the standard geologic timescale that began at the end of the Mesozoic era 65.5 million years ago and ended 10,000 years ago at the end of the Pleistocene epoch. Paleontologically referred to as "the age of mammals."

Ceratopsids/ceratopsians: These were large-bodied (13–26-foot long), quadrupedal, horned dinosaurs with highly varied and elaborate frills. A monophyletic group of ornithischians.

Clade: A lineage of genetically related organisms that has a common ancestor, in other words, a monophyletic group.

Cohort: A group of organisms that share a common way of making a living such as carnivory, herbivory, burrowing, etc. It can also designate a group of individuals in a population that are the same age.

Diamicts/diamictite: A terrigenous sedimentary rock that is nonsorted to poorly sorted and therefore contains a wide range of particle sizes. Examples can range from a pebbly sandstone to a cobble and boulder containing mudstone.

Dinosaur: The word literally means "terrible lizard." It is the common name for the Dinosauria, which is monophyletic and a well-supported clade that is included within the Archosauria clade. Reptiles are now considered a separate clade that doesn't include the Dinosauria. Dinosauria includes two major divisions or clades, Ornithischia and Saurischia, as well as Aves (birds). All members of the Dinosauria are terrestrial.

DNA: The acronym for deoxyribonucleic acid. It is a large molecule with a very high molecular weight that functions as the basic blueprint of heredity. DNA is a long-chain molecule comprised of nucleotides that are made up of three parts: phosphate, sugar, and a base. In DNA, the sugar is deoxyribose. The most commonly accepted model is that of Crick and Watson in which there are two strands that are linked by sugars and a base, and all, in turn, form a double helix or spiral.

Endotherms/ectotherms: Endotherms are animals that can regularly produce body heat and control their body temperatures internally. They do this usually within very narrow limits (homoeothermy). Ectotherms have little or no ability to internally produce and regulate their body temperatures and are much more subjected to the vagaries of ambient temperatures.

Epicontinental/epeiric seas: Ancient, extensive, shallow, interior seas that drowned continental interiors during high stands of global sea level.

Flood geologists: Individuals with degrees or some training in the geosciences who follow the teachings of George McCready Price that the Earth's sedimentary record was formed by a universal flood—Noah's flood of Genesis 6:9. These individuals are basically fundamentalist Christians who believe the Earth is young (six to ten thousand years old) and who adopt a literal reading of the Old Testament. They are often associated with the Creation Research Society.

Fossil: Any evidence of once living organisms that is at least prehistoric in age. Fossils are often permineralized, but they don't have to be. Fossils can retain biomolecules and soft parts such as tissues and can take the form of artifacts such as nests, traces of activity, such as tracks and trails, skeletal components, and/or molds and casts of same.

Hadrosaur: A major division or clade of ornithopod dinosaurs that were characterized by highly specialized tooth batteries in their jaws and flared predentary and premaxillary bones, which gave them a "duck-billed" appearance. There were two main types or clades, hadrosaurinae, which didn't have crests, and lambeosaurines, which had various types of crests and elaborate nasal passages. All were primarily bipedal but some were semiquadrupeds. All seem to have been herbivorous. Hadrosaurs first appeared in the Late Cretaceous (Turonian) over ninety million years ago but were most diverse and abundant in the latter part of the Late Cretaceous (Campanian and Maastrichtian) some sixty-five to eighty million years ago. This clade was the most widespread terrestrial nonavian animal on record.

Ichnites: A fossil footprint, track, or trail.

Laser: An acronym of "light amplification by stimulated emission of radiation." A electromagnetic wave (light) is produced by stimulating an electron with a photon. Laser light is most commonly a very narrow beam of "coherent light" that is monochromatic.

Mammoth: An extinct type of elephant (proboscidean) that belongs to the genus *Mammuthus.* This type of elephant had a unique dentition comprised of multiple enamel-covered and dentine-cored plates that were held together by cementum. Mammoths first appeared during the middle Pliocene epoch in sub-Saharan Africa and became extinct some eight to ten thousand years ago.

Megatrackway: Rock surfaces or layers covered by concentrations of footprints that cover geographic areas encompassing hundreds and thousands of square miles or kilometers. First defined by Martin Lockley.

North Slope: A term that is presently applied to most of Arctic Alaska. It is applied to an area in northern Alaska that extends from the axis of the Brooks Range to the shores of the Chukchi and Beaufort seas. This term originated with petroleum exploration geoscientists who undertook extensive fieldwork in this area during the early 1960s. Geomorphologists recognize three formal physiographic provinces within the area referred to as the North Slope. The Arctic Mountains, the Arctic Foothills, and the Arctic Coastal Plain.

Original horizontality: A fundamental principle or rule of stratigraphy that states that strata originally formed on a surface that was horizontal or close to horizontal. There are exceptions to this rule, but original horizontality of strata should be assumed unless there is evidence to the contrary.

Ornithopods: A stem-based taxon of small to large bipedal, unarmored, herbivorous dinosaurs. They are anatomically closer to hadrosaurids like *Edmontosaurus* than to ceratopsians like *Triceratops.* The clade includes heterodontosaurids, iguanodontids, and hadrosaurids. The ornithopods ranged stratigraphically from the Lower Jurassic to the uppermost Cretaceous.

Paleo-Arctic: At or above a paleolatitude of 66.5° N. A designation that is required because continents and ocean crust has shifted and changed position on the Earth's surface over time.

Paleoethology: The study of ancient animal behavior based on such things as closely associated organisms, tracks, traces, and artifacts such as nests.

Paleomagnetism: Ancient remnant magnetism that retains the original orientation (vectors) of the Earth's magnetic field at the time that the rock formed. The magnetism is retained in magnetically susceptible minerals such as magnetite.

Permafrost: Soil, sediment, and rock that has remained at or below the freezing point of water (32° F or 0° C) for at least two consecutive years. Permafrost is most common in polar regions but is also found in alpine areas or in the seabed even at low latitudes. Permafrost reaches depths of thousands of feet in the Arctic, where it is fairly continuous or contains subzones or patches of thawed ground.

Petroleum: Oil, gas, and related distillates that have been retained in rocks for some time.

Petroleum traps: Subsurface geologic structures that form spaces that confine oil and gas in rocks. These structures often involve folds, faults or the pinching out of a highly porous bed. In some cases, oil or gas will form in a highly porous bed that will be confined by beds of little or no porosity.

Pingo: An ice-cored hill or mound that is usually 90–150 feet (30–50 meters) high and up to 400 yards (400 meters) in diameter. The mound can be likened to an ice blister that formed as freezing water expanded and forced soil and sediment upward and outward. Most commonly found in Arctic regions.

Pleistocene: A geologic time division. It is an epoch of the Neogene period that ranges from 1.8 million years ago to 10,000 years ago and is often referred to as the Ice Age or glacial epoch.

Polymers/Polymerization: Huge, "super" molecules that are formed by repeated addition or condensation of large numbers of simple molecules. Examples are nucleic acids, proteins, nylon, resins, and Teflon.

Sedimentology: A study of the natural physical and chemical processes that form sediments and sedimentary deposits. It includes detailed studies of the environments of deposition and such things as primary structures, grain sizes, types of cement, and the physics of both dynamic and static fluids.

Sequence stratigraphy: A branch of stratigraphy that attempts to construct a time-stratigraphic framework based on cycles of worldwide changes of sea level. Packages of sedimentary rocks are divided using unconformities—subaerial erosion surfaces as well as flooding surfaces. The sea-level changes are induced by tectonic and climatic processes.

Silica: An abundant naturally occurring mineral with a basic formula of SiO_2. Kinds of silica are quartz, chalcedony, agate, and carnelian. All varieties are relatively hard. This mineral is a common cement that binds sediment and helps to form sedimentary rock.

Stem group: A clade that is defined by all the ancestral taxa that are related to one another. This is represented on a cladogram (a treelike diagram) by one entire main branch with all of its twigs or subdivisions. A clade is a monophyletic group—that is, a group that shares an ancestor species and all its descendant species.

Superposition: An important principle or rule of stratigraphy that is related to original horizontality. If there is no evidence to indicate otherwise, then whenever a stack of beds or strata are found, the oldest will be at the bottom and each successive bed will be younger than the one below it.

Taphonomy: A part of paleoecology that concentrates on the processes that produce fossils. This includes studies of natural processes and patterns of death, burial, and fossilization such as replacement and permineralization. This specialty overlaps with studies of sediment and sedimentary rock formation.

Taxon: A group of organisms that are assigned a proper name using a recognized code such at the Linnaean system. A taxon can be a species, a genus, a family, or any higher rank.

Theropods/Theropoda: A monophyletic clade of meat-eating saurischian dinosaurs that typically had serrated teeth, hollow bones, and large openings in front of the orbits in the skull. There is still quite a bit of debate regarding the clades, or lineages, that should be included within the theropods. Most specialists agree that Theropoda should include allosaurs, ceratosaurs, troodontids, dromaeosaurs, oviraptorids, and tyrannosaurs.

Track: A fossil imprint or footprint. A general term applied to this type of trace fossil is ichnite. Some specialists use "track" as a synonym for "trackway."

Trackway: A sequence of fossil footprints or tracks.

Uniformitarianism: "The present is a key to the past." A basic principle of modern geology that relies on fundamental biologic, physical, and chemical laws and processes that are operating today and that underlies the interpretation of rocks and rock formations. It holds that physical laws have not changed through time but have only varied in respect to rates and combinations. The principle does not preclude catastrophes but it does emphasize the dominant role of gradualism-uniformity throughout geologic history.

NOTES

1. The Arctic Setting

1. The fossil horsetail rush *Equisetites* is abundantly represented by root casts in what were once sediments depositing in ponds, along river banks, and in floodplain wetlands during the Cretaceous. The modern equivalent of this taxon is *Equisetum*, popularly known as the scouring or horsetail rush. This plant incorporates abundant amounts of microcrystals of silica in its cell walls, making it highly abrasive on teeth. This and plants such as ferns and cycads also contain silica bodies called phytolithsTrueplant stones." All of these plants were abundant when hadrosaurs and ceratopsians roamed Alaska's paleo-Arctic. These highly abrasive plant tissues may have been a key factor in the evolution of the tooth batteries that marked both of these lineages of dinosaurs during the Late Cretaceous.

2. Prasad, Strömberg, Habib, et al. (2005). It shouldn't surprise anyone working on the paleoecology of the Late Cretaceous that several varieties of grasses had evolved by this time. At present, we have no direct evidence of grasses in the paleo-Arctic record. However, comprehensive phytolith sampling of hadrosaur and ceratopsian teeth has not been conducted as yet, but it is a promising research direction. The first angiosperms date back to the Early Cretaceous (Aptian) and were rapidly evolving by the Campanian. See Fastovsky and Smith (2004), 619, for an excellent graphic summary of the Cretaceous plant record.

3. Totman Parrish and Spicer (1988); Spicer, Totman Parrish, and Grant (1992); Spicer (2003).

4. Nikiforuk (1984); Varricchio, Jackson, Scherzeer, et al. (2005).

5. Velikovsky (1955), 13–15. Also see Goldsmith (1977) for discussions that question Velikovsky's logic and his grasp of astronomy and planetary dynamics. Although this collection of essays concentrates on Velikovsky's *Worlds in Collision*, there are a number of refutations of Velikovsky's assumptions and basic premises that are pertinent to his *Earth in Upheaval*.

6. The Arctic Monitoring and Assessment Program (AMAP 1997; 1998) and Conservation of Arctic Flora and Fauna (CAFF 2001) are the sources of mapped information used in plate 1. The Arctic is defined by latitude, vegetation, and temperature.

7. Williams (2003); Berton (2001). Also see Roderick (1997), 135.

8. Lopez (2001).

9. *The Thing from Another World* is a 1951 movie that takes place at a research base in Alaska. An alien "killer vegetable" is found frozen in ice, having crashed there on a journey from outer space. In the 1978 movie, Superman is compelled by the power of kryptonite to fly to the Arctic Circle to discover who he really is and where he came from.

10. Michener (1988); Davis, Liston, and Whitmore (1998).

11. See figure 1.3.

12. Cohen (2002).

13. *The Lost World* is the title of a book written by Sir Arthur Conan Doyle of Sherlock Holmes fame. In *The Lost World*, Doyle tells the tale of an expedition to a mysterious plateau in the Amazon Basin of South America that has preserved a world of the past with all manner of ancient critters, including dinosaurs. Published in 1912, the book was the basis for a Hollywood movie in 1925 and influenced the plots of several movies and books that followed, including Michael Crichton's *The Lost World*, published in 1995.

14. Cohen (2002), 42.

15. Cohen (2002), 61.

16. Norman (1985); Sarjeant (1997).

17. Rudwick (1985).

18. Sagan (1996).

2. Tracks Lead the Way

The epigraph is from McPhee (1998), 64. In his *Annals of the Former World*, McPhee, a journalist, writes a book that celebrates the science of geology in a way that is almost poetic. More importantly, he writes in a way that communicates clearly and excitedly how geologists piece together the fascinating history of the earth. He undertakes a series of journeys across the United States during which he engages with a series of active geologic researchers from various backgrounds and geologic specialties, who he portrays as human beings living out their passion for discovery and understanding.

1. McFarling and Lafferty (2003); Fookes (2008), 150; Berkman and Young (2009).

2. Scotese, Gahagan, Ross, et al. (1987); Smith, Hurley, and Briden (1981); Rich , Gangloff, and Hammer (1997), 564.
3. Lapparent (1962).
4. Lapparent (1962), 18.
5. Hurum, Milàn, Hammer, et al. (2006).
6. Edwards, Edwards, and Colbert (1978).
7. Upchurch, Otto-Bliesner, and Scotese (1999).
8. Rich, Gangloff, and Hammer (1997), 563–64.
9. Núñez-Betelu, Hills, and Currie (1995); Chen, Osadetz, Embry, et al. (2000). See also plate 8.
10. Grady (1993), 155.
11. Núñez-Betelu, Hills, and Currie (2005).
12. Vandermark, Tarduno, and Brinkman, et al. (2009).
13. Russell (1984).
14. David Evans (personal communication, 2010).
15. Yorath and Cook (1981).
16. Smith, Hurley, and Briden (1981); Rich, Gangloff, and Hammer (1997).
17. David Evans (personal communication, 2010).
18. Smith, Hurley, and Briden (1981); Rich, Gangloff, and Hammer (1997).
19. Gangloff, May, and Storer (2004).
20. Cameron and Beaton (2000); Lowey (2010).
21. Suslov (1961).
22. Davies (1983); Fiorillo (2008a); Main and Scotese (2009).
23. Rich, Gangloff, and Hammer (1997), 563.
24. Kolosov, Ivensen, Mikhailova, et al. (2009).
25. Kolosov (2008). Verified by Pascal Godefroit (personal communication, 2010).
26. Rozhdestvenskiy (1973); Rich, Gangloff, and Hammer (1997).
27. Averianov, Voronkevich, Maschenko, et al. (2002).
28. Kurzanov, Efimov, and Gubin (2000).
29. Nessov (1995); Golovneva (1994) figure 1, 90.
30. Golovneva (1994); Godefroit, Golovneva, Shchepetov, et al. (2009).
31. Godefroit (personal communication, 2010).

3. A Black Gold Rush Sets the Stage for Discovery in Alaska

1. Dickinson (1988) contains a nice summary of the long-standing controversy within the geoscience community concerning the origin of petroleum. It is written for the nonspecialist and general reader. Haun (1974), a collection of papers that first appeared in the *American Association of Petroleum Geologists Bulletin*, provides a highly technical discussion of the possible origins of petroleum.
2. Roderick (1997) is a fairly accurate and very readable book on the history of oil development in Alaska. The author characterizes his book as one that is more about the politics of oil than the science. However, Roderick does a fine job of recounting the contributions of some of the geologists and land managers to this history. Roderick reminds the reader of how critical geoscience and geoscientists were to the greatest petroleum discovery in North America and to the bank account that the state of Alaska now shares with all of its citizens through the Permanent Fund.
3. Roderick (1997), chapters 9 and 10.
4. Wahrhaftig (1965).
5. Rich, Gangloff, and Hammer (1997) and (2002).
6. U. S. Census Bureau (2010)
7. McGinniss (1980).
8. Gates (2006).
9. Gangloff (2003).
10. Roderick (1997); Smith (1993).
11. Clemens (1994); Gangloff and Fiorillo (2010), appendix 1, 317. The bone bed that contained the dinosaur remains that Robert Liscomb collected in 1961 named the Liscomb Bone Bed. It has been formally defined as a rock unit by Gangloff and Fiorillo (2010).
12. Gangloff (2003). The poem was found among Robert Liscomb's belongings after his death and provided through the kindness of his sister Mrs. Bonnie Liscomb Brundage. Robert's name was not found on the poem, but it was most probably written by him. I have in part changed the order of the original.
13. Davies (1987).
14. Brouwers, Clemens, Spicer, et al. (1987).
15. Alvarez, Alvarez, Asaro, et al. (1980).
16. Clemens (1982); Clemens, Archibald, and Hickey (1981). Ward (1992), chapter 5, describes Ward's biostratigraphic work on ammonites along the spectacular cliffs in Zumaya, Spain. Ward discusses the K-T extinctions from the perspective of the marine fossil record.
17. A bolide is a meteor that explodes in the earth's atmosphere rather than impacting the earth's surface. Some have argued that the extraterrestrial impactor that caused the extinction of the dinosaurs and their kin was possibly a comet. See Raup (1991), 156–66. Raup does a great job of discussing all of the great extinctions found in the geologic record, including the K-T extinction. He does this in clear, concise language, tinged with humor that is meant to reach the general reader.
18. Bakker (1980); Chinsamy (1990).
19. For a detailed and comprehensive look at the dinosaur extinction debate, see Archibald (1996); Russell and Dodson (1997); Archibald and Fastovsky (2004).
20. Turco, Toon, Ackerman, et al. (1983). The term "nuclear winter" did not originate with the Alverez, Alvarez, Asaro, et al. (1980) paper. The term was first employed by Turco and his associates in 1983.

21. Farlow (1990); Padian and Horner (2004).
22. Totman Parrish and Spicer (1988); Spicer, Totman Parrish, and Grant (1992).
23. Norman (1985); Sarjeant (1997)
24. Alvarez and Asaro (1990) is a less technical discussion of the impact scenario with respect to dinosaur extinction at the end of the Cretaceous than the Alvarez, Alvarez, Asaro, et al. (1980) paper. This second article expands the original hypothesis and includes evidence that was not presented in the original paper.
25. Angier (1985).
26. Davies (1987).
27. Brouwers, Clemens, Spicer, et al. (1987), 1608.
28. Parrish, Totman Parrish, Hutchinson, et al. (1987).
29. Letters section, *Science* 239, 10–11.
30. Paul (1988).
31. Currie (1989).
32. Angier (1985).
33. The label "creationist" is essentially equivalent to that of biblical fundamentalist. Creation scientists can usually be distinguished from other fundamentalists in that they often hold degrees in engineering or other scientific disciplines and attempt to reconcile scientific facts with Old Testament literalism. Flood geologists can be distinguished from other creation scientists by their focus on reconciling the "flood of Noah," in the Book of Genesis, with data and techniques derived from the geosciences. The majority of flood geologists either hold degrees in the geosciences from secular universities and colleges or are students in geoscience programs at creationists colleges.
34. See http://www.creationontheweb.org/content/view/4007 for this description of Margaret Helder along with a short bio and a listing of seven articles that she has written for *Creation Magazine* (accessed Jan. 2009).
35. Davies (1987), 198.
36. Brouwers, Clemens, Spicer, et al. (1987), 1608.
37. Goodwin, Grant, Bench, et al. (2007).
38. Gangloff and Fiorillo (2010).
39. Helder (1992).
40. Helder (2003). See http://www.freenet.edmonton.ab.ca/articles/pachyrh.html (accessed Jan. 2009).
41. "Cheechako" is a popular epithet that Alaskans apply to newcomers. It is roughly equivalent to the label of "greenhorn" as used in the lower forty-eight.
42. Davis, Liston, and Whitmore (1998).

4. Peregrines, Permafrost, and Bone Beds

1. A pingo is one of the few positive topographic features on Arctic coastal plains. It is a small rounded hill with an ice core that can be seen for tens of miles on a clear day. See the glossary for more details regarding the formation of pingos.
2. Orth (1967). The Colville River is primarily formed by Thunder and Storm creeks in the De Long Mountains and flows some 350 east then northeast until it forms a large delta at the head of Harrison Bay along the Arctic Ocean coast. The river has been given several aboriginal (Eskimo) names meaning "goose," "headwaters," or "big river."
3. Witte, Stone, and Mull (1987); Brouwers, Clemens, Spicer, et al. (1987); Phillips (2003). The Liscomb Bone Bed site has been determined, using paleomagnetic techniques, to have had a paleolatitude as high as 85° N.
4. Rogers and Kidwell (2007), 3–5; Eberth, Shannon, and Nolan (2007).
5. Conrad, McKee, and Turin (1990); Phillips (2003).
6. Raup (1991).
7. Rich and Vickers-Rich (2000).
8. Gangloff and Fiorillo (2010), figure 8, 306 and figures 10 A, B, 307.
9. The minerals silica and calcite are the most common natural cements that form in sedimentary rocks; they fill openings in the bone and replace original bone minerals and internal structures as well as cement the sediment particles together to form rock. See the glossary in this book for further details on these two minerals.
10. Fiorillo and Gangloff (2001), figure 3, 327; Gangloff and Fiorillo (2010), 303–305, and figure 9, 307.
11. "Trackway" as used here refers to a series of tracks or footprints of a single individual bird or dinosaur. See the glossary for more details.
12. Mull, Houseknecht, and Bird (2003). This publication represents the latest conclusions regarding the Upper Cretaceous rock packages found on Alaska's North Slope. It is the first major revision of the stratigraphic nomenclature that was followed for some fifty years by most field geologists and paleontologists. See figure 4.16 for the revised nomenclature and a schematic representation of the rock packages.
13. Gangloff (1995).
14. Horner (1979). This was the first extensive compilation of dinosaur remains that were found in marine sediments in North America. This record combined with finds in the Talkeetna Mountains of Alaska and subsequent finds in Baja, California, all point to a close relationship between Late Cretaceous dinosaurs and the extensive shorelines that marked the paleogeography of North America during the last part of the Cretaceous time period. One possible explanation for this close association relates to the possibility that dinosaurs were taking advantage of rich accumulations

of marine foods, such as seaweed and sea grasses. Another is that conditions for preservation were more favorable than those present in more inland areas.

15. Pasch and May (2001).

16. Conyers (1978).

17. Parrish, Totman Parrish, Hutchinson, et al. (1987), figure 5, 381.

18. "Articulation" here refers to finding bones in contact that functioned together or articulated when the animal was alive.

19. Fiorillo and Gangloff (2001); Gangloff and Fiorillo (2010).

20. Parrish, Totman Parrish, Hutchison, et al. (1987).

21. Gangloff (1998b).

22. See the glossary.

23. Lockley (1991), 215.

24. Gangloff (1998b).

25. O. J. Smith was one of the cadre of Alaskan bush pilots who gained their wings as combat pilots in World War II and the Korean conflict. They often transitioned into bush flying after stints as airline pilots. O.J., as he liked to be called, flew for Wien Air Alaska before setting up shop at Umiat, on Alaska's North Slope, where he made use of a motley collection of old vehicles, trailers, and various other buildings left over from state and federal programs that used Umiat as a base over the years. Although most people that ran into him remember him as the gray-haired, wiry, and plain-speaking character that anointed himself as the "mayor of Umiat," he had a distinguished career as a young C-47 pilot during the Normandy invasion. He and his two sons left their indelible marks on the North Slope. O. J. holds a special place in my memory book along with Walt Audi, who served as our principal bush pilot and who was a contemporary of O. J. Smith.

26. Barringer (2005). This journalist has done a fine job of describing the present importance of Umiat, as well as its climate and some of its colorful history.

27. Barringer (2005).

28. Fiorillo, McCarthy, Brandlen, et al. (2010).

29. From 1993 to 1999 my field and laboratory research on Alaska's dinosaurs was awarded significant grants from Title II Dwight D. Eisenhower Fund, the Dinosaur Society Research Fund, the University of Alaska President's Special Projects Fund, and the Jurassic Foundation.

30. Gangloff (1997).

31. Rich, Gangloff, and Hammer (1997).

5. Texas, Teachers, and Chinooks

1. "Oil patch" presently refers to any region that is known for its production of oil and natural gas. Originally, the label was applied to the petroleum-rich midcontinent oil region that extended from central Texas across Oklahoma to eastern Kansas. This region became a world-class petroleum producing region in the early 1920s. Texas and Oklahoma became synonymous with oil between the 1920s and the end of World War II. However, California and other western states developed their "oil patches" during this same period. After World War II, a second oil patch that included the Gulf Coast of Texas, Louisiana, and Mississippi was developed. By the 1970s, Alaska's North Slope had become North America's greatest single producer of petroleum and part of a new "oil patch" that extends from the Cook Inlet of south-central Alaska across northern Alaska to Canada's Mackenzie River delta and then south through Alberta. Many of the petroleum geologists, oil field workers, and oil company managers learned their trade in the midcontinent region before heading for Alaska. Houston, Texas, is still considered by many in the oil business to be the "capital" of the U.S. petroleum industry.

2. Fiorillo and Totman Parrish (2004); Fiorillo, Kucinski, and Hamon (2004); Fiorillo (2006).

3. Gangloff (1997).

4. Sorensen (2000).

5. Like many Alaskan place names, Deadhorse elicits smirks and inquiries as to the origin of such a moniker, and, like other place names, there are a variety of tall or bizarre stories attached to it. The most plausible is that the airport received its name from a contractor that hauled gravel during its construction. Google the origin for some light entertainment if you have nothing better to do.

6. A Brunton compass is a standard geological field instrument. It is a magnetic compass combined with an inclinometer. It is used to take the cross-country trend and inclination of bedded rocks and to make geologic maps.

7. Senkowsky (2003); Fiorillo (2004).

8. Stratigraphy is the study of the sequences of layers of sedimentary and some volcanic rocks, while sedimentology focuses on the physics, chemistry, and processes that produce layered rocks. Sequence stratigraphy is the study of the patterns of deposition of sedimentary rocks, which reflect major and minor changes in these patterns, as they relate to worldwide and local sea-level changes. The detailed patterns can be used to correlate depositional and erosional events from one area to another and therefore to establish a time framework.

9. Fiorillo, Tykoski, Currie, et al. (2009).

10. Fiorillo and Gangloff (2000).

11. May and Gangloff (1999); Gangloff (1999); Fiorillo, McCarthy, Brandlen, et al. (2010); Gangloff and Fiorillo (2010).

12. Fiorillo, McCarthy, Brandlen, et al. (2010), 473.

13. Spicer and Parrish (1990); Godefroit, Golovneva, Shchepetov, et al. (2008); Fiorillo, McCarthy, Brandlen, et al. (2010).

14. Strahler and Strahler (1991).

15. Ridgway, Trop, and Sweet (1997)

16. Smith, Hurley, and Briden (1981); Fiorillo, Hasiotis, Kobayashi, et al. (2009). The Cantwell Formation was deposited in a very complex plate tectonic setting in which two major plates, the North American and Wrangellia, were colliding. To complicate this aspect further, the upper Cantwell Paleogene volcanics have been determined to have originated at a paleolatitude as high as 83° N (Hillhouse and Grommé 1982).

17. Fiorillo, Hasiotis, Kobayashi, et al. (2009).

18. Fister (2006).

19. Kevin May (oral communication, 2002). May has tentatively identified dinosaur ichnites but has not conducted a detailed comprehensive study of this find as yet. Kevin is a highly experienced fossil vertebrate tracker and has a very keen eye when it comes to recognizing fossil tracks and trackways. He was with me when we found the first dinosaur tracks and trackways in the Yukon Territory of Canada (Gangloff, May, and Storer 2004).

20. Hotton (1980); Parrish Totman Parrish, Hutchinson, et al. (1987); Currie (1989).

21. The term "endothermy" refers to the ability of an animal to produce and regulate its body heat internally and to maintain its temperature within narrow limits (homeothermy). This physiologic attribute is often erroneously referred to as "warm-bloodedness" by the general public. A living reptile or amphibian cannot regulate its temperature internally and is subject to the vagaries of the ambient temperature in its environment. It may overheat if it is caught in certain circumstances and die due to "hot-bloodedness."

22. Paul (1988), Paul (1994).

23. Fiorillo and Gangloff (2001), 328; Bell and Snively (2008), 272.

24. Bell and Snively (2008) have presented an excellent recent analysis and review of the status of the dinosaur migration debate.

25. Lehman (2001).

26. Fricke, Rogers, and Gates (2009).

27. May and Gangloff (1999); Fiorillo and Gangloff (2000); Gangloff, Fiorillo, and Norton (2005); Fiorillo and Gangloff (2010); Fiorillo, McCarthy, Brandlen, et al. (2010).

28. Godefroit, Golovneva, Shchepetov, et al. (2008).

29. Fanti and Miyashita (2009).

30. Carpenter and Alf (1994); Mark Goodwin (personal communication and University of California Museum of Paleontology specimen V99143/169072).

31. Grady (1993).

32. Paul (1994), 406; Fiorillo and Gangloff (2001).

33. Gangloff (1995).

34. Fiorillo and Gangloff (2000); Gangloff. Fiorillo, and Norton (2005); Bell and Snively (2008); Godefroit, Golovneva, Shchepetov, et al. (2008).

35. Parrish, Totman Parrish, Hutchison, et al. (1987); Currie (1989).

36. Fancy, Pank, Whitten, et al. (1989); Fiorillo and Gangloff (2000); Paul (1997); Bell and Snively (2008).

37. Fancy, Pank, Whitten, et al. (1989), 645, 646; Valkenburg (1998).

38. Paul (1994).

39. Geist (1998); Gangloff and Fiorillo (2010), 314.

40. Bergerud (1985); Nadon (1993).

41. Totman Parrish and Spicer (1988); Rich, Vickers-Rich, and Gangloff (2002); Spicer (2003).

42. Geist (1998), 331.

43. Varricchio, Martin, and Katsura (2007).

44. Rich and Vickers-Rich (2000), figures 24 B–C, 57, and chapter 11.

45. Fiorillo and Gangloff (2000); Norman, Sues, Witmer, et al. (2004); Godefroit, Golovneva, Shchepetov, et al. (2008); Fiorillo, Tykoski, Currie, et al. (2009).

46. Lockley and Hunt (1995).

47. Lockley (1991).

6. The Arctic during the Cretaceous

1. Water gains and loses heat energy slower than does rock and soil that makes up the land surfaces. Water is said to have a higher specific heat than rock and soil, and it requires about five times the amount of heat energy to raise its temperature to the same degree in comparison to dry rock and soil. Therefore during the winter, land masses lose heat energy more quickly and cool more rapidly compared to water bodies (lakes and oceans). The opposite relationship is seen during the summer period—land masses heat up faster than do water bodies. This contrast in the concentration of heat energy between land masses and water bodies is also responsible for the development of air pressure, which, in turn, controls the movement of air in the lower atmosphere. High pressure forms over cold surfaces, while lower pressures form over warm surfaces. Air moves from higher to lower pressure concentrations. For an excellent and more detailed discussion of these important phenomena, see Strahler and Strahler (1991).

2. Williams and Stelck (1975); Hay, DeConto, Wold, et al. (1999); Kump and Slingerland (1999).

3. Kauffman (1984); Hay, DeConto, Wold, et al. (1999).

4. See Dott and Batten (1981). This is a highly popular college textbook for introductory historical geology courses. Chapters 15 and 16 are particularly

cogent. In 1994, a seventh edition, authored by Donald Prothero Jr. and Robert Dott, was published in which the emphasis was changed to the evolution of life. For a much more technical and up-to-date discussion of Cretaceous paleogeography, see Barrera and Johnson (1999).

5. Totman Parrish (1998).

6. Totman Parrish and Spicer 1988; Spicer, Totman Parrish, and Grant (1992); Rich, Vickers-Rich, and Gangloff (2002); Spicer (2003); Spicer and Herman (2010).

7. Barrera and Savin (1999).

8. Parrish, Totman Parrish, Hutchinson, et al. (1987), 377, 378; Brouwers, Clemens, Spicer, et al. (1987); Godefroit, Golovneva, Shchepetov, et al. (2008).

9. See Wyllie (1976). Despite its age, this is still the best introduction to the basics of plate tectonics that I have come across in my forty years of teaching geoscience. It is profusely and clearly illustrated, and Wyllie does a superb job of reviewing the relationship between the paradigms of continental drift, seafloor spreading, and plate tectonics. The book was written at a key point in the history of the debate concerning the validity and relationships of these paradigms. Van Andel (1985) is a more recent attempt to elucidate continental drift and the theory of plate tectonics. Van Andel's perspective is more ocean oriented and has sections that focus on the ways life and climate were affected by the transient ocean floor and continental plates as they evolved.

10. Spicer, Totman Parrish, and Grant (1992), figure 4, 185.

11. McCabe and Totman Parrish (1992).

12. The North Slope of Alaska is particularly rich in Cretaceous coals and is estimated to have more than half of the total U.S. reserves. See Smith (1993) and Flores, Stricker, and Scott (2003).

13. Horner (1979); Paul (1988); Gangloff and Fiorillo (2010).

14. Phillips (2003), 518–20, and figure 17B.

15. Larkum and den Hartog (1989).

16. Bergerud (1985); Nadon (1993).

17. Epicontinental and epeiric are shallow seas that form over continental crust as opposed to seas and oceans that form over deep ocean basins. Such seas are typically warmer and develop circulation patterns and weather that are almost independent of those that are found in the ocean basins at the same time. See Kump and Slingerland (1999).

18. Horner (1979).

19. Phillips (2003); Lehman (2001); Eberth (2010); Fiorillo, McCarthy, Brandlen, et al. (2010).

20. Butler and Barrett (2008); Eberth (2010).

21. Pasch and May (2001).

22. Boaz (1982); Nikiforuk (1984); Weigelt (1989); Capaldo and Peters (1995); Varricchio, Jackson, Scherzeer, et al. (2005); Fiorillo, McCarthy, Brandlen, et al. (2010).

23. Currie and Dodson (1984); Varricchio and Horner (1993); Gangloff and Fiorillo (2007); Ralrick and Tanke (2008); Fiorillo, McCarthy, Brandlen, et al. (2010).

24. These percentages were compiled from chapters 3, 5, 13, 17, 20, and 23 in Weishampel, Dodson, and Osmólska (2004). It is also interesting to see the percentage (76 percent) of complete or nearly complete skeletons from the Late Cretaceous of Asia and North America that are attributed to hadrosaurs and ceratopsians in Russell and Bonaparte (1997). Also see Lockley (1997), 573, for the proportions of ornithopods versus theropod tracks in the best-studied Late Cretaceous sites.

25. Nelms (1989); Gangloff and Fiorillo (2010).

26. Gangloff (1998b); Currie, Langston, and Tanke (2008); Fiorillo, McCarthy, Brandlen, et al. (2010).

27. Dodson, Forster, and Sampson (2004); Lehman (2001); Sampson and Loewen (2010).

28. Parrish, Totman Parrish, Hutchison, et al. (1987). An occipital condyle is the ball joint at the back of the skull that fits into and articulates with the first neck vertebra. The assignment to the subfamily Chasmosaurinae Lambe, 1915, is based primarily on the shape of an isolated bone, and the attribution to the genus *Anchiceratops* is very tentative. The horn cores discussed by Parrish et al. are even more problematic for their taxonomic assignment.

29. Dodson, Forster, and Sampson (2004).

30. Dodson (1996).

31. Forster (1997); Fastovsky and Smith (2004).

32. Tiffney (1997). Be sure to see his discussion of the Late Cretaceous, 366–68.

33. Dodson (1996), 264–65; Williams, Barrett, and Purnell (2009).

34. Eberth (2010).

35. Norman, Sues, Witmer, et al. (2004).

36. Gangloff, Fiorillo, and Norton (2005); Godefroit, Golovneva, Shchepetov, et al. (2008).

37. Vickaryous, Maryańska, and Weishampel (2004)

38. Gangloff (1995).

39. Godefroit, Golovneva, Shchepetov, et al. (2008).

40. Maryańska, Chapman, and Weishampel (2004).

41. Clemens (1994); Norman, Sues, Witmer, et al. (2004).

42. Godefroit, Golovneva, Shchepetov, et al. (2008).

43. Norman, Sues, Witmer, et al. (2004), figures 18.5 and 18.6.

44. Troyer (1984); Townsend, Slapcinsky, Krysko, et al. (2005).

45. Xiao-chun and Sues (1996).

46. Fiorillo (1991); Chure, Fiorillo, and Jacobsen (1998).

47. Fiorillo (1991).

48. Jacobsen (1998).

49. Fiorillo and Gangloff (2000); Fiorillo (2008b). *Troodon* teeth from the Late Cretaceous of Arctic Alaska are, on the average, twice the size of those recorded for this taxon from lower latitudes. Larger denticles and teeth would have provided greater prey-holding power and might have been an evolutionary adaptation that developed in a predator that had other advantages over larger carnivores, as pointed out in this chapter. This attribute could have allowed Arctic *Troodon* to take on larger prey than its populations farther south.

50. Varricchio (1997); Currie and Dong (2001).

51. Varricchio (1997), 749; Norell and Makovicky (2004). The greatest variety and most complete remains of troodontids are from the famous Late Cretaceous formations of Mongolia and northwestern China. Much of this evidence was accrued during Sino-Canadian expeditions between 1987 and 1990.

52. Longrich (2010) has noted the anatomical evidence for relatively large eyes in the ceratopsian *Protoceratops*. An eye size to body mass ratio comparison using data from extant birds led him to conclude that large eyes evolved as the result of several life styles in birds. Chief among these are predation, scavenging, and nocturnal habits. When Longrich applied this anatomical ratio to a range of dinosaur taxa, he found that *Protoceratops* was the closest taxon to some of the largest-eyed extant birds of any of the dinosaurs in his dataset. Unfortunately, he did not include *Troodon* in his dinosaur dataset.

53. Gangloff and Fiorillo (2010).

54. Varricchio, Martin, and Katsura (2007). Longrich (2010) cites evidence for possible denning in *Protoceratops*. If Longrich is correct, this would represent evolution of this strategy in two different clades of herbivorous dinosaurs that lived in very different paleoenvironments.

55. Ryan, Currie, Gardner, et al. (1998).

56. Fiorillo and Gangloff (2000).

57. Fiorillo and Gangloff (2000), table 1, 677.

58. Guggisberg (1975), esp. 282–83. I have observed this predator-prey feeding displacement behavior between lions, hyenas, and cheetahs in the Masai Mara of southwestern Kenya. Extrapolating behavior in living animals to fossil forms is a branch of paleoecology called paleoethology and is a type of uniformitarianism called actualism. The study of animal behavior in fossil animals is a fascinating but very nascent branch of paleontology and evolutionary biology. Food sources and pathways can be identified using stable isotopes and trace elements. Social behavior can be inferred using trackway patterns. Detailed age profiles from bone bed assemblages can point to rudimentary social structure and behavior. Nesting grounds can reveal patterns and skeletal associations that point to nursery organization and egg-laying patterns and that can indicate whether the newly born were precocial or altricial.

7. Cretaceous Dinosaur Pathways in the Paleo-Arctic and along the Western Interior Seaway

1. Lockley (1991). Dodson, Forster, and Sampson (2004) note that there is a consistent pattern of ceratopsids and ocean shorelines. A recent study by Johnson, Ledesma-Vazquez, and B. Gudveig Baarli (2006) in Baja, California, Mexico report dinosaur skeletal material and tracks being found along a Cretaceous rocky shoreline.

2. Johnson, Ledesma-Vazquez, and Gudveig (2006), 152; Lockley and Hunt (1995).The first use of the term "megatracksites" is ascribed to Lockley and Pittman (1989).

3. Lockley (1991), 194.

4. Lockley (1991), 125–30.

5. Rylaarsdam, Varban, Plint, et al. (2006).

6. Cyanobacteria or so-called blue-green algae and green algae form strings of interlocking cells that form microscopic mats under the right conditions of moisture, pH, and abundant sunshine. These mats will then "armor" the sediment surfaces so that currents will not erase tracks and other surface impressions of vertebrate and invertebrate animals. Under the proper conditions, even sand dune surfaces can preserve animal tracks as can be seen in Triassic and Jurassic age rocks in the southwestern United States.

7. Roehler and Stricker (1984).

8. Currie and Sarjeant (1979).

9. The major and lesser mountain ranges that are part of the Western Cordillera of North America experienced their first orogenies (volcanism, magmatism, and folding) during the Early Mesozoic era. The time of greatest deformation and magmatism was from the Late Jurassic through Early Paleogene (195 to 60 million years ago).

10. Lockley (1991; 1995) was the first in North America to use the metaphor "freeway" for relatively narrow dinosaur trackways that extend for hundreds of miles along shorelines and mountain fronts.

11. McCrea and Currie (1998); McCrea, Lockley, and Meyer (2001).

12. Lockley (1991).

13. An ichnogenus is a formal taxonomic designation applied to the tracks of vertebrate and invertebrate animals. It should be considered a form-genus that is not equivalent to a genus name formally applied within the Linnean binomial taxonomic system.

14. Rylaarsdam, Varban, Plint, et al. (2006).
15. Gangloff, May, and Storer (2004).
16. Stott (1975); Stott (1984).
17. Currie (1989); McCrea, Lockley, and Meyer (2001).
18. Currie and Sarjeant (1979); Currie (1983).
19. Gradstein, Ogg, and Smith (2004).
20. Davies (1983); Kirkland (2005).
21. Gangloff, May, and Storer (2004).
22. The Old World is defined as the continents of Australia and Eurasia that make up the Eastern Hemisphere.
23. Iseman (1988).
24. Hopkins (1982); Main and Scotese (2009).
25. Davies (1983); Milner and Norman (1984).
26. Gangloff (1998b); Godefroit, Golovneva, Shchepetov, et al. (2008); Fiorillo (2008a).
27. Godefroit, Bolotsky, and Van Itterbeeck (2004); Kirkland (2005).
28. Lillegraven, Kielan-Jaworowska, and Clemens (1979); Matthews (1998); Poinar (2005).
29. Godefroit, Bolotsky, and Van Itterbeeck (2004), 613.
30. Davies (1983); Chinnery-Allgeier and Kirkland (2010); Main and Scotese (2009)
31. Rich and Vickers-Rich (2000); Varricchio, Martin, and Katsura (2007).
32. Roehler and Stricker (1984); Fiorillo and Totman Parrish (2004); Gangloff, May, and Storer (2004).

8. Applying New Technologies to the Ancient Past

1. All instruments were optically based when I took my undergraduate civil engineering course in surveying. Laser-based instruments and GPS are rapidly replacing optical instruments such as alidades and plane tables to construct detailed topographic maps of fossil sites. See the glossary for a more complete definition of the word "laser."
2. Balanoff and Rowe (2007).
3. Witmer and Ridgely (2008).
4. Permafrost is not only found in the polar regions. It underlies the higher elevations of mountains in temperate and tropical zones and can be found in the seafloor in such places as the Gulf of Mexico. During the summer, an active melted zone develops from the surface down to a few feet (a meter or so). This causes heaving and loss of cohesiveness that can result in road breakup, collapse of buildings, and mud and debris slides (Suslov 1961; Strahler and Strahler 1991; Conner and O'Haire 1988). See the glossary for a further definition.
5. Rich and Vickers-Rich (2000).
6. Rich and Vickers-Rich (2000), 186–88.
7. The majority of the gold that has been mined in the Fairbanks area and other parts of interior Alaska has come from depressions or placer pockets eroded out of the bedrock. Placer gold is rounded and worn, as it has been eroded from bedrock veins and transported by ancient rivers to where it is now found.
8. Abbott (2007); Rich (2008). These two sources provide accounts that are quite different in their emphasis and also with respect to what transpired during the excavation of the tunnel in March and April 2007. The Abbott story is from the point of view of a fairly naïve TV producer who raised the money to make a documentary about what she thought was an exciting new approach to digging dinosaurs, and Arctic dinosaurs at that. Unfortunately, the mix of personalities and the rigors of work on the North Slope produced a tunnel, but it also produced some bitterness and a reality lesson for the producer. I have had several experiences with documentaries, film crews, and producers while I was working on the North Slope over my eighteen years of dinosaur pursuit. Most of these documentaries turned out to be unpleasant, time consuming, and of questionable worth. I found that most film crews were ill prepared to deal with the rigors and demands of the conditions in the Arctic, even in summer. They were prone to make unrealistic demands of me, my crews, and our logistical staff. I will never forget an incident in 1996 when a film crew asked our bush pilot to taxi his aircraft and then take off four times so they could get the "right" sequence. The scene was not in the final product and must have ended up in the editors "wastebasket"—I couldn't bring myself to tell the pilot, and, fortunately, he never saw the finished video.
9. Fiorillo, McCarthy, Brandlen, et al. (2010).
10. Poinar and Poinar (1994), 204; Schweitzer (1997); Miller, Drautz, Ratan, et al. (2008).
11. Yang (1997).
12. Pääbo (1993); Cohen (2002).
13. Morell (1993); Woodward, Weyland, and Bunnell (1994).
14. Schweitzer, Johnson, Zocco, et al. (1997).

9. Natural Resources, Climate Change, and Arctic Dinosaurs

1. Sanborn (1976). Uranium salts (highly oxidized and hydrated UO_2) are water-soluble forms of uranium minerals that have been found to concentrate in petrified wood and dinosaur bone as well as the surrounding fluvial sediments. Carnotite is one of the more commonly known minerals that is made up of uranium salts. See Health Physics Society answer 1952 at http://hps.org.

2. Barnes (1967); Gates (2006). Latest statistics on sources and consumption of energy in the United States can be found at http://eia.doe.gov.

3. Flores, Stricker, and Scott (2003).

4. Roehler and Stricker (1984); Gangloff (1998a). The regularity with which dinosaur fossils are found in coal deposits has been pointed out and documented in chapters 4 and 5 of the present book.

5. Poinar and Poinar (1994).

6. Poinar and Poinar (1994), 102–103.

7. Langenheim, Smiler, and Gray (1960).

8. Ahlbrandt, Huffman, Fox, et al. (1979); Roehler and Stricker (1984).

9. Poinar and Poinar (1994).

10. Poinar and Poinar (1994), 195–99, offer an answer that is far more eloquent and informative than I am capable of providing. In 2008, a team of twenty-one researchers reported that they had reconstructed perhaps 50 percent of the genome of the Pleistocene-age woolly mammoth by recovering the DNA from the hair of a twenty-thousand- and sixty-five-thousand-year-old individual exhumed from permafrost in Siberia. So, we have taken a big step toward "Jurassic Park." See Miller, Drautz, Ratan, et al. (2008).

11. Petroleum includes oil, natural gas, and related distillates.

12. Chen, Osadetz, Embry, et al. (2000); Kleinberg and Brewer (2001); Klett, Bird, Brown, et al. (2007).

13. Main and Scotese (2009).

14. Berkman and Young (2009); Parfitt (2009).

15. Norton (2003).

10. Future Expansion of the Arctic Dinosaur Record

1. Currie and Koppelhaus (2005).

2. Ryan and Russell (2001); Tanke (2007).

3. Lehman (2001); Ryan and Russell (2001).

4. Fiorillo and Gangloff (2000).

5. Chen, Osadetz, Embry, et al. (2000).

6. Russell (1984); Rich, Gangloff, and Hammer (1997).

7. Yorath and Cook (1981); Feinstein, Brooks, Fowler, et al. (1988).

8. Cameron and Beaton (2000); Lowey (2010).

9. Gangloff (1997); Gangloff (1998a); Gangloff (1998b).

10. Huffman (1989); Harris, Mull, Reifenstuhl, et al. (2002).

11. Mull, Houseknecht, and Bird (2003).

12. McLellan, Servheen, and Huber (2008).

13. Conyers (1978). Theropod and ornithopod tracks are reported in Weishampel(1990), 86. However, Weishampel, Barrett, Coria, et al. (2004), 545, report only theropod tracks from this locality. My investigations of original photos and descriptions of the Black Lake site have led me to agree with the latter. See figure 2.1 of the present book.

14. Fiorillo and Totman Parrish (2004).

15. Godefroit, Golovneva, Shchepetov, et al. (2008)

16. Previously designated the K-T or Cretaceous-Tertiary extinction, the Tertiary period is no longer a valid geological time term and has been replaced by the Paleogene (Gradstein, Ogg, and Smith, 2004).

17. Rich , Gangloff, and Hammer (1997); Averianov, Voronkevich, Maschenko, et al. (2002); Kurzanov, Efimov, and Gubin (2003).

18. Weishampel, Barret, Coria, et al. (2004).

19. Rozhdestvenskiy (1973); Lauters, Bolotsky, Van Itterbeeck, et al. (2008).

20. Godefroit, Bolotsky, and Van Itterbeeck (2004); Van Itterbeeck, Bolotsky, Bultynck, et al. (2005).

21. Smith, Hurley, and Briden (1981).

22. Shinkarev (1973); Kontorovich (2007).

LITERATURE CITED

Abbott, A. 2007. Tunnel Vision. *Nature* 450, 18–20.

Ahlbrandt, T. S., A. C. Huffman Jr., J. E. Fox, and I. Pasternack. 1979. Depositional Framework and Reservoir-Quality Studies of Selected Nanushuk Group Outcrops, North Slope, Alaska. In T. S. Ahlbrandt, ed., *Preliminary Geologic, Petrologic, and Paleontologic Results of the Study of Nanushuk Group Rocks, North Slope, Alaska*, 14–32. U. S. Geological Survey Circular 794. Washington, DC: GPO.

Alexander, R. McN. 1989. *Dynamics of Dinosaurs and Other Extinct Giants*. New York: Columbia University Press.

———. 1997. Engineering a Dinosaur. In J. O. Farlow and M. K. Brett-Surman, eds., *The Complete Dinosaur*, 414–25. Bloomington: Indiana University Press.

Alvarez, L. W., W. Alvarez, F. Asaro, and H. V. Michel. 1980. Extraterrestrial Cause for the Cretaceous-Tertiary Extinction. *Science* 208, 1095–108.

Alvarez, L. W., and F. Asaro. 1990. An Extraterrestrial Impact. *Scientific American* 263, 78–84.

Angier, N. 1985. Did Comets Kill the Dinosaurs? A Bold New Theory About Mass Extinctions. *Time* 125, 72–83.

Arctic Monitoring and Assessment Programme [AMAP]. 1997 and 1998. *Assessment Report: Arctic Pollution Issues*. Oslo, Norway. http://www.amap.no.

Archibald, J. D. 1996. *Dinosaur Extinction and the End of an Era*. New York: Columbia University Press.

Archibald, J. D., and D. E. Fastovsky. 2004. Dinosaur Extinction. In D. B. Weishampel, P. Dodson, and H. Osmólska, eds., *The Dinosauria*, 2nd ed., 672–84. Berkeley: University of California Press.

Averianov, A. O., A. V. Voronkevich, E. N. Maschenko, S. V. Leshchinskiy, and A. V. Fayngertz. 2002. A Sauropod Foot from the Early Cretaceous of Western Siberia, Russia. *Acta Palaeontologica Polonica* 47, 117–24.

Bakker, R. T. 1980. Dinosaur Heresy-Dinosaur Renaissance. In R. D. K. Thomas and E. C. Olson, eds., *A Cold Look at the Warm-Blooded Dinosaurs*, 351–462. American Association for the Advancement of Science Selected Symposium 28. Boulder, CO: Westview Press.

Balanoff, A. M., and T. Rowe. 2007. Osteological Description of an Embryonic Skeleton of the Extinct Elephant Bird, *Aepyornis* (Palaeognathae: Ratitae). Society of Vertebrate Paleontology Memoir 9, supplement 4, 1–53

Barnes, F. F. 1967. *Coal Resources of Alaska*. U.S. Geological Survey Bulletin 1242-B. Washington, DC: GPO.

Barrera, E., and C. C. Johnson, eds., 1999. *Evolution of the Cretaceous Ocean-Climate System*. Geological Society of America Special Paper 332. Boulder, CO: Geological Society of America.

Barrera, E., and S. M. Savin. 1999. Evolution of Late Campanian-Maastrichtian Marine Climates and Oceans. In E. Barrera and C. C. Johnson, eds., *Evolution of the Cretaceous Ocean-Climate System*, 245–82. Geological Society of America Special Paper 332. Boulder, CO: Geological Society of America.

Barringer, F. 2005. Coldest Place in the U. S. Is a Hotspot in Its Own Tiny Way. *New York Times News Service*, August 19.

Bell, P. B., and E. Snively. 2008. Polar Dinosaurs on Parade: A Review of Dinosaur Migration. *Alcheringa* 32, 271–84.

Bergerud, A. T. 1985. Antipredator Strategies of Caribou: Dispersion along Shorelines. *Canadian Journal of Zoology* 63, 1324–29.

Berkman, P. A., and O. R. Young. 2009. Governance and Environmental Change in the Arctic Ocean, *Science* 324, 339–40.

Berton, P. 2001. *Arctic Grail: The Quest for the Northwest Passage and the North Pole, 1818–1909*. Toronto, Canada: Anchor.

Boaz, D. 1982. Modern Riverine Taphonomy: Its Relevance to the Interpretation of Plio-Pleistocene Hominid Paleoecology in the Omo Basin, Ethiopia. PhD diss., University of California, Berkeley.

Brouwers, E. M., W. A. Clemens, R. A. Spicer, T. A. Ager, L. D. Carter, and W. V. Sliter. 1987. Dinosaurs on the North Slope, Alaska: High Latitude, Latest Cretaceous Environments. *Science* 237, 1608–10.

Butler, R. J., and P. M. Barrett. 2008. Palaeoenvironmental Controls on the Distribution of Cretaceous Herbivorous Dinosaurs. *Naturwissenschaften* 95, 1027–32.

Cameron, A. R., and A. P. Beaton. 2000. Coal Resources of Northern Canada with Emphasis on Whitehorse Trough, Bonnet Plume Basin and Brackett Basin. *Journal of International Coal Geology* 43, 187–210.

Capaldo, S. D., and C. R. Peters. 1995. Skeletal Inventories from Wildebeest Drownings at Lakes Masek and Ndutu in the Serengeti Ecosystem of Tanzania. *Journal of Archaeological Science* 22, 385–408.

Carpenter, K. 1999. *Eggs, Nests, and Baby Dinosaurs: A Look at Dinosaur Reproduction.* Bloomington: Indiana University Press.

Carpenter, K., and K. Alf. 1994. Global Distribution of Dinosaur Eggs, Nests, and Babies. In K. Carpenter, K. F. Hirsch, J. R. and Horner, eds., *Dinosaur Eggs and Babies*, 15–30. New York: Cambridge University Press.

Chen, Z., K. G. Osadetz, A. F. Embry, H. Gao, and P. K. Hannigan. 2000. Petroleum Potential in Western Sverdrup Basin, Canadian Arctic Archipelago. *Bulletin of Canadian Petroleum Geology* 48, 323–38.

Chinnery-Allgeier, B. J., and J. I. Kirkland. 2010. An Update on the Paleobiogeography of Ceratopsian Dinosaurs. In M. J. Ryan, B. J. Chinnery-Allgeier, and D. A. Eberth, eds., *New Perspectives on Horned Dinosaurs: The Royal Tyrrell Museum Ceratopsian Symposium*, 387–404. Bloomington: Indiana University Press.

Chinsamy, A. 1990. Physiological Implications of the Bone Histology of *Syntarsus rhodesiensis* (Saurischia: Theropoda). *Paleontologia Africana* 27, 77–82.

Chure, D. J., A. R. Fiorillo, and A. Jacobsen, A. 1998. Prey Bone Utilization by Predatory Dinosaurs in the Late Jurassic of North America, with Comments on Prey Bone Use by Dinosaurs throughout the Mesozoic. *GAIA* 15, 227–32.

Clemens, W. A. 1982. Patterns of Extinction and Survival of the Terrestrial Biota during the Cretaceous/Tertiary Transition. Geological Society of America Special Paper 190, 407–13. Boulder, CO: Geological Society of America.

———. 1994. Continental Vertebrates from the Late Cretaceous of the North Slope, Alaska. In D. K. Thurston and K. Fujita, eds., *1992 Proceedings International Conference on Arctic Margins.* Outer Continental Shelf-Mineral Management Service Publication 94-0040, 395–98. Anchorage, AK: U. S. Department of the Interior.

Clemens, W. A., J. D. Archibald, and L. J. Hickey. 1981. Out with a Whimper Not a Bang. *Paleobiology* 7, 293–98.

Clemens, W. A., and L. G. Nelms. 1993. Paleoecological Implications of Alaskan Terrestrial Vertebrate Fauna in Latest Cretaceous Time at High Paleolatitudes. *Geology* 21, 503–506.

Cohen, C. 2002. *The Fate of the Mammoth: Fossils, Myth, and History.* Chicago: University of Chicago Press.

Conner, C., and D. O'Haire. 1988. *Roadside Geology of Alaska.* Missoula, MT: Mountain Press Publishing Company.

Conrad, J. E., E. H. McKee, and B. D. Turrin. 1990. *Age of Tephra Beds at the Ocean Point Dinosaur Locality, North Slope, Alaska, Based on K-Ar and*^{40}*Ar/*^{39}*Ar Analyses.* U. S. Geological Survey Bulletin 1990-C, 1–12. Washington, DC: GPO.

Conservation of Arctic Flora and Fauna [CAFF]. 2001. *Status and Conservation.* Akureyri, Iceland. http://www.CAFF.is.html.

Conyers, L. 1978. Letters, Notes, and Comments. *Alaska Magazine* 44, 30.

Crichton, M. 1995. *The Lost World.* New York: Knopf.

Currie, P. J. 1983. Hadrosaur Trackways from the Lower Cretaceous of Canada. *Acta Palaeontologica Polonica* 28, 63–73.

———. 1989. Long Distance Dinosaurs. *Natural History* 89, 59–65.

Currie, P. J., and P. Dodson. 1984. Mass Death of a Herd of Ceratopsian Dinosaurs. In W. E. Reif and F. Westphal, eds., *Third Symposium on Mesozoic Terrestrial Ecosystems*, 61–66. Tubingen: Attempto.

Currie, P. J., and Z.-M Dong. 2001. New Information on Cretaceous Troodontids (Dinosauria, Theropoda) from the People's Republic of China. *Canadian Journal of Earth Science* 38, 1753–66.

Currie, P. J., and E. B. Koppelhaus, eds., 2005. *Dinosaur Provincial Park: A Spectacular Ancient Ecosystem Revealed.* Bloomington: Indiana University Press.

Currie, P. J., W. Langston Jr., and D. H. Tanke. 2008. *A New Horned Dinosaur from an Upper Cretaceous Bone Bed in Alberta.* Ottawa: NRC Research Press.

Currie, P. J., and W. A. S. Sarjeant. 1979. Lower Cretaceous Dinosaur Footprints from the Peace River Canyon, British Columbia, Canada. *Palaeogeography, Palaeoclimatology, Palaeoecology* 28, 103–15.

———. 1989. Dinosaur Footprints of Western Canada. In D. D. Gillette and M. G. Lockley, eds., *Dinosaur Tracks and Traces,* 293–300. Cambridge: Cambridge University Press.

Davies, K. L. 1983. Hadrosaurian Dinosaurs of Big Bend National Park Brewster County, Texas. MA thesis, University of Texas at Austin.

———. 1987. Duck-Bill Dinosaurs (Hadrosauridae, Ornithischia) from the North Slope of Alaska. *Journal of Paleontology* 61, 198–200.

Davis, B., M. Liston, and J. Whitmore. 1998. *The Great Alaskan Dinosaur Adventure: A Real-Life Journey through the Frozen Past.* Green Forest, AR: Master Books.

Dickinson, W. W. 1988. Plankton to Petroleum. *Earth Science,* Winter, 21–23.

Dodson, P. 1996. *The Horned Dinosaurs A Natural History.* Princeton, NJ: Princeton University Press.

Dodson, P., C. A. Forster, and S. D. Sampson. 2004. Ceratopsidae. In D. B. Weishampel, P. Dodson, and H. Osmólska, eds., *The Dinosauria,* 2nd ed., 494–513. Berkeley: University of California Press.

Dott, R. H., Jr., and R. L. Batten. 1981. *Evolution of the Earth.* New York: McGraw-Hill.

Eberth, D. A. 2010. A Review of Ceratopsian Paleoenvironmental Associations and Taphonomy. In M. J. Ryan, B. J. Chinnery-Allgeier, and D. A. Eberth, eds., *New Perspectives on Horned Dinosaurs: The Royal Tyrrell Museum Ceratopsian Symposium,* 428–46. Bloomington: Indiana University Press.

Eberth, D. A., M. Shannon, and B. G. Nolan. 2007. A Bonebeds Database: Classification, Biases, and Patterns of Occurrence. In R. R. Rogers, D. A. Eberth, and A. R. Fiorillo, eds., *Bonebeds: Genesis, Analysis, and Paleobiological Significance,* 103–320. Chicago: University of Chicago Press.

Edwards, M. B., R. Edwards, and E. H. Colbert. 1978. Carnosaurian Footprints in the Lower Cretaceous of Eastern Spitzbergen. *Journal of Paleontology* 52, 940–41.

Fancy, S. G., L. F. Pank, K. R. Whitten, and W. L. Regelin. 1989. Seasonal Movements of Caribou in Arctic Alaska as Determined by Satellite. *Canadian Journal of Zoology* 67, 644–50.

Fanti, F., and T. Miyashita. 2009. A High Latitude Vertebrate Fossil Assemblage from the Late Cretaceous of West-Central Alberta, Canada: Evidence for Dinosaur Nesting and Vertebrate Latitudinal Gradient. *Palaeogeography, Palaeoclimatology, Palaeoecology* 275, 37–53.

Farlow, J. O. 1990. Dinosaur Energetics and Thermal Biology. In D. B. Weishampel, P. Dodson, and H. Osmólska, eds., *The Dinosauria,* 43–55. Berkeley: University of California Press.

Fastovsky, D. E., and J. B. Smith. 2004. Dinosaur Paleoecology. In D. B. Weishampel, P. Dodson, and H. Osmólska, eds., *The Dinosauria,* 2nd ed., 614–26. Berkeley: University of California Press.

Feinstein, S., P. W. Brooks, M. G. Fowler, L. R. Snowdon, and G. K. Williams. 1988. Families of Oils and Source Rocks in the Central Mackenzie Corridor: A Geochemical Oil-Oil and Oil-Source Rock Correlation. In D. P. James and D. A. Leckie, eds., *Sequences, Stratigraphy, Sedimentology: Surface and Subsurface,* 543–52. Canadian Society of Petroleum Geologists Memoir 15. Alberta: Canadian Society of Petroleum Geologists.

Fiorillo, A. R. 1991. Prey Bone Utilization by Predatory Dinosaurs. *Palaeogeography, Palaeoclimatology, Palaeoecology* 88, 157–66.

———. 2004. The Dinosaurs of Arctic Alaska. *Scientific American* 291, 85–91.

———. 2006. Review of the Dinosaur Record of Alaska with Comments Regarding Korean Dinosaurs as Comparable High-Latitude Fossil Faunas. *Journal of Paleontology of South Korea* 22, 15–27.

———. 2008a. *Dinosaurs of Alaska: Implications for the Cretaceous Origin of Beringia.* Geological Society of America Special Paper 442, 313–26. Boulder, CO: Geological Society of America.

———. 2008b. On the Occurrence of Exceptionally Large Teeth of *Troodon* (Dinosauria: Saurischia) from the Late Cretaceous of Northern Alaska. *Palaios* 23, 322–28.

Fiorillo, A. R., and R. A. Gangloff. 2000. Theropod Teeth from the Prince Creek Formation (Cretaceous) of Northern Alaska, with Speculations on Arctic Dinosaur Paleoecology. *Journal of Vertebrate Paleontology* 20, 675–82.

———. 2001. The Caribou Migration Model for Arctic Hadrosaurs (Dinosauria: Ornithischia): A Reassessment. *Historical Biology* 15, 323–34.

Fiorillo, A. R., S. T. Hasiotis, Y. Kobayashi, and C. S. Tomsich. 2009. A Pterosaur Manus Track from Denali National Park, Alaska Range, Alaska, United States. *Palaios* 24, 466–72.

Fiorillo, A. R., R. Kucinski, and T. R. Hamon. 2004. New Frontiers, Old Fossils: Recent Dinosaur Discoveries in Alaska National Parks. *Alaska Park Science* 3, 4–9.

Fiorillo, A. R., P. J. McCarthy, E. Brandlen, P. P. Flaig, D. Norton, L. Jacobs, P. Zippi, and R. A. Gangloff. 2010. Paleontology and Paleoenvironmental Interpretation of the Kikak-Tegoseak Quarry (Prince Creek Formation: Late Cretaceous), Northern Alaska: A Multi-Disciplinary Study of a High-Latitude

Ceratopsian Dinosaur Bonebed. In M. J. Ryan, B. J. Chinnery-Allgeier, and D. A. Eberth, eds., *New Perspectives on Horned Dinosaurs: The Royal Tyrrell Museum Ceratopsian Symposium*, 456–77. Bloomington and Indianapolis: Indiana University Press.

Fiorillo, A. R., and J. Totman Parrish. 2004. The First Record of a Cretaceous Dinosaur from Southwestern Alaska. *Cretaceous Research* 25, 453–58.

Fiorillo, A. R., R. S. Tykoski, P. J. Currie, P. J. McCarthy, and P. Flaig. 2009. Description of Two Partial *Troodon* Braincases from the Prince Creek Formation (Upper Cretaceous), North Slope, Alaska. *Journal of Vertebrate Paleontology* 29, 178–87.

Fister, K. 2006. Discovery of New Dinosaur Evidence in Denali National Park and Preserve. *Denali National Park and Preserve News*, 1–2. http://www.nps.gov//dena/parknews/dinosaurdiscovery.htm.

Flores, R. M., G. D. Stricker, and A. K. Scott. 2003. *Alaska Coal Geology; Resources, and Coalbed Methane Potential.* U.S. Geological Survey Bulletin 2198. http://pubs.usgs.gov/bul/b2198.

Fookes, P. G. 2008. Some Aspects of the Geology of Svalbard. *Geology Today* 24, 146–52.

Forster, C. A. 1997. Hadrosauridae. In P. J. Currie and K. Padian, eds., *Encyclopedia of Dinosaurs*, 295–99. San Diego, CA: Academic Press.

Fricke, H. C., R. R. Rogers, and T. A. Gates. 2009. Hadrosaurid Migration: Inferences based on Stable Isotope Comparisons Among Late Cretaceous Dinosaur Localities. *Paleobiology* 3, 270–88.

Gangloff, R. A. 1995. *Edmontonia* sp., the First Record of an Ankylosaur from Alaska. *Journal of Vertebrate Paleontology* 15, 195–200.

———. 1997. The Arctic Alaska Dinosaur Program. *Arctic Research of the United States* 11, 50–54.

———. 1998a. Paleontological and Archaeological Research in the Eastern Third of the National Petroleum Reserve-Alaska: A Call for Symbiosis. In J. E. Martin, J. W. Hoganson, and R. C. Benton, eds., *Proceedings of the Partners Preserving Our Past, Planning Our Future*, 79–84. Dakoterra 5. Rapid City: South Dakota School of Mines and Technology.

———. 1998b. Arctic Dinosaurs with Emphasis on the Cretaceous Record of Alaska and the Eurasian-North American Connection. In S. G. Lucas, J. I. Kirkland, and J. W. Estep, eds., *Lower and Middle Cretaceous Terrestrial Ecosystems*, 211–20. New Mexico Museum of Natural History and Science Bulletin 14. Albuquerque: New Mexico Museum of Natural History and Science.

———. 2003. Alaska, a Crucible for Science. *Polar Times* 3, 3–4.

Gangloff, R. A., and A. R. Fiorillo. 2007. Taphonomy and Paleoecology of a Remarkably Rich Upper Cretaceous High Latitude Bonebed from the Prince Creek Formation, North Slope Alaska. *Journal of Vertebrate Paleontology* 27, supplement 3, 79A.

———. 2010. Taphonomy and Paleoecology of a Bonebed from the Prince Creek Formation, North Slope, Alaska. *Palaios* 25, 299–317.

Gangloff, R. A., A. R. Fiorillo, and D. Norton. 1999. Arctic Dinosaurs and Their Cretaceous Record in Alaska. 50th Arctic Science Conference, Arctic Division of the American Association for the Advancement of Science, *Program and Abstracts*, 288.

———. 2005. The First Pachycephalosaurine (Dinosauria) from the Paleo-Arctic of Alaska and its Paleogeographic Implications. *Journal of Paleontology* 79, 997–1001.

Gangloff, R. A., K. C. May, and J. E. Storer. 2004. An Early Late Cretaceous Dinosaur Tracksite in Central Yukon Territory, Canada. *Ichnos* 11, 299–309.

Gates, N., ed. 2006. *The Alaska Almanac.* 30th ed. Anchorage: Alaska Northwest Books.

Geist, V. 1998. *Deer of the World: Their Evolution, Behavior, and Ecology.* Mechanicsburg, PA: Stackpole Books.

Godefroit, P., Y. L. Bolotsky, and J. Van Itterbeeck. 2004. The Lameosaurine Dinosaur *Amurosaurus riabinini*, from the Massstrichtian of Far Eastern Russia. *Acta Palaeontologica Polonica* 49, 585–618.

Godefroit, P., L. Golovneva, S. Shchepetov, G. Garcia, and P. Alekseev. 2008. The Last Polar Dinosaurs: High Diversity of Latest Cretaceous Arctic Dinosaurs in Russia. *Naturwissenschaften* 96, 495–501.

Goldsmith, D. 1977. *Scientists Confront Velikovsky.* Ithaca, NY: Cornell University Press.

Golovneva, L. 1994. The Flora of the Maastrichtian-Danian Deposits of the Koryak Upland, Northeastern Russia. *Cretaceous Research* 15, 89–100.

Goodwin, M. B., P. G. Grant, G. Bench, and P. A. Holroyd. 2007. Elemental Composition and Diagenetic Alteration of Dinosaur Bone: Distinguishing Micron-Scale Spatial and Compositional Heterogeneity Using PIXE. *Palaeogeography, Palaeoclimatology, Palaeoecology* 253, 458–76.

Gradstein, F. M., J. G. Ogg, and A. G. Smith, eds., 2004. *A Geologic Time Scale 2004*. New York: Cambridge University Press.

Grady, W. 1993. *The Dinosaur Project: The Story of the Greatest Dinosaur Expedition Ever Mounted.* Toronto: Macfarlane, Walter and Ross.

Guggisberg, C. A. W. 1975. *Wild Cats of the World.* New York: Taplinger.

Harris, E. E., C. G. Mull, R. R. Reifenstuhl, and S. Montayne. 2002. *Geologic Map of the Dalton Highway (Atigun Gorge to Slope Mountain) Area, Southern Arctic Foothills, Alaska.* PIR 2002-2. Fairbanks, AK: Division of Geological and Geophysical Surveys.

Haun, J. D., ed. 1974. *Origin of Petroleum II.* AAPG Reprint Series no. 9. Tulsa, OK: American Association of Petroleum Geologists.

Hay, W. H., R. M. DeConto, C. N. Wold, K. M. Wilson, S. Voigt, M. Schultz, A. Wold-Rossby, W.-C. Dullo, A. B. Ronov, A. N. Balukhovsky, and E. Söding. 1999. Alternative Global Cretaceous Paleogeography. In E. Barrera and C. C. Johnson, eds., *Evolution of the Cretaceous Ocean-Climate System,* 1–47. Geological Society of America Special Paper 332. Boulder, CO: Geological Society of America.

Haynes, G. 1991. *Mammoths, Mastodonts, and Elephants: Biology, Behavior, and the Fossil Record.* New York: Cambridge University Press.

Helder, M. 1992. Fresh Dinosaur Bones Found. *Creation Ex Nihilo* 14, 16–17.

———. 2003. The Big Splash. http://www.create.ab.ca/the-big-splash.

Hillhouse, J. W., and C. S. Grommé. 1982. Limits to Northward Drift of the Paleocene Cantwell Formation, Central Alaska. *Geology* 10, 552–56.

Hopkins, D. M. 1982. Aspects of the Paleogeography of Beringia during the Late Pleistocene. In D. M. Hopkins, J. V. Matthews Jr., C. E. Schweger, and S. B. Young, eds., *Paleoecology of Beringia,* 3–28. New York: Academic Press.

Horner, J. R. 1979. Upper Cretaceous Dinosaurs from the Bearpaw Shale (Marine) of South-Central Montana with a Checklist of Upper Cretaceous Dinosaur Remains from Marine Sediments in North America. *Journal of Paleontology* 53, 566–78.

Hotton, N. 1980. An Alternative to Dinosaur Endothermy. In R. D. K. Thomas and E. C. Olson, eds., *A Cold Look at the Warm-Blooded Dinosaurs,* 311–50. Boulder, CO: Westview.

Huffman, A. C., Jr. 1985. *Geology of the Nanushuk Group and Related Rocks, North Slope, Alaska.* U. S. Geological Survey Bulletin 1614. Washington, DC: GPO.

———. 1989. Nanushuk Group. In C. G. Mull and K. Adams, eds., *Bedrock Geology of the Eastern Koyukuk Basin, Central Brooks Range, and Eastcentral Arctic Slope along the Dalton Highway, Yukon River to Prudhoe Bay, Alaska,* 303–309. Guidebook 7, vol. 2. Anchorage, AK: Division of Geological and Geophysical Surveys.

Hurum, J. H., J. Milàn, Ø. Hammer, I. Midtkandal, H. Amundsen, and B. Sæther. 2006. Tracking Polar Dinosaurs: New Finds from the Lower Cretaceous of Svalbard. *Norwegian Journal of Geology* 86, 397–402.

Iseman, P. 1988. Lifting the Ice Curtain. *New York Times,* October 23.

Jacobsen, A. R. 1998. Feeding Behavior of Carnivorous Dinosaurs as Determined by Tooth Marks on Dinosaur Bones. *Historical Biology* 13, 17–26.

Johnson, M. E., J. Ledesma-Vazquez, and B. Gudveig Baarli. 2006. Vertebrate Remains on Ancient Rocky Shores: A Review with Report on Hadrosaur Bones from the Upper Cretaceous of Baja California (Mexico). *Journal of Coastal Research* 22, 574–80

Kauffman, E. G. 1984. Paleobiogeography and Evolutionary Response Dynamic in the Cretaceous Western Interior Seaway of North America. In G. E. G. Westermann, ed., *Jurassic-Cretaceous Biochronology and Palaeogeography of North America,* 273–306. Geological Association of Canada Special Paper 27. St. Johns, Canada: Geological Association of Canada.

Kirkland, J. 2005. Utah's Newly Recognized Dinosaur Record from the Early Cretaceous Cedar Mountain Formation. *Utah Geological Survey Notes* 37, 1–5.

Kleinberg, R. L., and P. G. Brewer. 2001. Probing Gas Hydrate Deposits. *American Scientist* 89, 244–51.

Klett, T. R., K. J. Bird, P. J. Brown, R. R. Carpentier, D. L. Gautier, D. W. Houseknecht, T. E. Moore, J. K. Pitman, R. W. Saltud, C. J. Schenk, and M. E. Tennyson. 2007. *Assessment of Undiscovered Petroleum Resources of the Laptev Sea Shelf Province, Russian Federation.* U.S. Geological Survey Fact Sheet 3096.

Kolosov, P. N. 2008. Dinosaurs of the Asian North-East. *Dinosaur Publications.* . http://www.dinosaursociety.com/asian_pub.php.

Kolosov, P. N., G. V. Ivensen, T. E. Mikhailova, S. M. Kurzanov, M. B. Efimov, and Yu. M. Gubin. 2009. Taphonomy of the Upper Mesozoic Tetrapod Teete Locality (Yakutia). *Paleontology Journal* 43, 201–207.

Kontorovich, A. 2007. Eastern Prospects of Russia's Oil Industry. *Oil of Russia* 2. http://www.oilru.com/or/31/581/.

Kump, L. R., and R. L. Slingerland. 1999. Circulation and Stratification of the Early Turonian Western Interior Seaway: Sensitivity to a Variety of Forcings. In E. Barrera and C. C. Johnson, eds., *Evolution of the Cretaceous Ocean-Climate System*, 181–90. Geological Society of America Special Paper 332. Boulder, CO: Geological Society of America.

Kurzanov, S. M., M. B. Efimov, and Yu. M. Gubin. 2000. Dinosaurs of Yakutia. In A. V. Komarov, ed., *Regional Conference on Geological Materials from Siberia, Eastern and Far Eastern Russia* (in Russian), vol. 2, 356–57. Tomsk: Gala Press.

———. 2001. Jurassic Dinosaurs of Yakutia. *Journal of Vertebrate Paleontology* 20, supplement 3, 70A.

———. 2003. New Archosaurs from the Jurassic of Siberia and Mongolia. *Paleontology Journal* 37, 53–57.

Lambe, L. M. 1920. *The Hadrosaur Edmontosaurus from the Upper Cretaceous of Alberta*. Geological Survey of Canada Memoir 120, 1–79.

Langenheim, R. L., Jr., C. J. Smiler, and J. Gray. 1960. Cretaceous Amber from the Arctic Coastal Plain of Alaska. *Geological Society of America Bulletin* 71, 1345–56.

Lapparent, A. F. de. 1960. The Discovery of Dinosaurian Footprints in the Cretaceous of Spitzbergen. *Comptes rendu des l'Academie des sciences* 251, 1399–1400.

———. 1962. Footprints of Dinosaur in the Lower Cretaceous of Vestspitsbergen-Svalbard. *Norsk Polarinstitute Årbok*, 14–21.

Larkum, A. W. D., and C. den Hartog. 1989. Evolution and Biogeography of Seagrasses. In A. W. D. Larkum, A. J. McComb, and S. A. Shepherd, eds., *Biology of Seagrasses: A Treatise on the Biology of Seagrasses with Special Reference to the Australian Region*, 112–55. New York: Elsevier.

Lauters, P., Y. L. Bolotsky, J. Van Itterbeeck, and P. Godefroit. 2008. Taphonomy and Age Profile of a Latest Cretaceous Dinosaur Bone Bed in Far Eastern Russia. *Palaios* 23, 153–62.

Lehman, T. M. 2001. Late Cretaceous Dinosaur Provinciality. In D. H. Tanke and K. Carpenter, eds., *Mesozoic Vertebrate Life*, 310–28. Bloomington: Indiana University Press.

Lillegraven, J. A., Z. Kielan-Jaworowska, and W. A. Clemens. 1979. Mesozoic Mammals: the First Two-Thirds of Mammalian History. In J. A. Lillegraven, M. J. Kraus, and T. M. Bown, eds., *Paleogeography of the World of the Mesozoic*, 294–96. Berkeley: University of California Press.

Lockley, M. G. 1991. *Tracking Dinosaurs: A New Look at an Ancient World.* New York: Cambridge University Press.

———. 1997. The Paleoecological and Paleoenvironmental Utility of Dinosaur Tracks. In J. O. Farlow and M. K. Brett-Surman, eds., *The Complete Dinosaur*, 554–78. Bloomington: Indiana University Press.

Lockley, M. G., and A. P. Hunt. 1995. *Dinosaur Tracks and Other Fossil Footprints of the Western United States.* New York: Columbia University Press.

Lockley, M. G., and J. G. Pittman. 1989. The Megatracksite Phenomenon: Implications for Paleoecology, Evolution and Stratigraphy. *Journal of Vertebrate Paleontology* 9, supplement 3, 30A.

Longrich, N. 2010. The Function of Large Eyes in Protoceratops: A Nocturnal Ceratopsian? In M. J. Ryan, B. J. Chinnery-Allgeier, and D. A. Eberth, eds., *New Perspectives on Horned Dinosaurs: The Royal Tyrrell Museum Ceratopsia Symposium*, 308–27. Bloomington: Indiana University Press.

Lopez, B. 2001. *Arctic Dreams: Imagination and Desire in a Northern Landscape*. New York: Vintage.

Lowey, G. W. 2010. The Bonnet Plume Basin, Yukon, Canada: Previously Unrecognized Oil Play. *Search and Discovery* 10228, 1–8.

Lull, R. S., and N. E. Wright. 1942. *Hadrosaur Dinosaurs of North America*. Geological Society of America Special Papers 40, 1–242. Boulder, CO: Geological Society of America.

Main, D. J., and C. R. Scotese. 2009. Paleobiogeographic Pathways: Beringia and Barentsia as Dinosaur Biogeographic Highways. *Geological Society of America Abstracts with Programs* 41, 31.Mandrik, I. 2010. Geological Exploration: Number One Priority. *Oil of Russia* 4. http://www.oilru.com/or/45/931/.

Maryańska, T., R. E. Chapman, and D. B. Weishampel. 2004. Pachycephalosauria. In D. B. Weishampel, P. Dodson, and H. Osmólska, eds., *The Dinosauria*, 2nd ed., 464–77. Berkeley: University of California Press.

Matthews, J. W. 1998. *Patterns of Freshwater Fish Ecology.*, Norwell, MA: Kluwer.

May, K. C., and R. A. Gangloff. 1999. New Dinosaur Bone Bed from the Prince Creek Formation, Colville River, National Petroleum Reserve-Alaska. *Journal of Vertebrate Paleontology* 19, supplement 3, 62A.

McCabe, P. J., and J. Totman Parrish. 1992. Tectonic and Climatic Controls on the Distribution and

Quality of Cretaceous Coals. In P. J. McCabe and J. Totman Parrish, eds., *Controls on the Distribution and Quality of Cretaceous Coals*, 1–15. Geological Society of America Special Paper 267. Boulder, CO: Geological Society of America.

McCrea, R. T., and P. J. Currie. 1998. A Preliminary Report on Dinosaur Tracksites in the Lower Cretaceous (Albian) Gates Formation Near Grande Cache, Alberta. In S. G. Lucas and J. W. Estep, eds., *Lower and Middle Cretaceous Terrestrial Ecosystems*, 155–62. New Mexico Museum of Natural History and Science Bulletin 14. Albuquerque: New Mexico Museum of Natural History and Science.

McCrea, R. T., M. G. Lockley, and C. A. Meyer. 2001. Global Distribution of Purported Ankylosaur Track Occurrences. In K. Carpenter, ed., *The Armored Dinosaurs*, 413–54. Bloomington: Indiana University Press.

McFarling, U. L., and M. Lafferty. 2003. Ice Shelf in Arctic Breaking into Pieces. *Polar Times* 3, 20.

McGinniss, J. 1980. *Going to Extremes*. New York: Knopf.

McLellan, B. N., C. Servheen, and D. Huber. 2008. *Ursus arctos*. In 2008 *International Union for Conservation of Nature Red List of Threatened Species*. http://www.iucn.org/about/work/programmes/species/red_list.

McPhee, J. 1998. *Basin and Range*. Vol. 1 of *Annals of the Former World*. New York: Farrar, Straus and Giroux.

Michener, J. A. 1988. *Alaska*. New York: Random House.

Miller, W., D. I. Drautz, A. Ratan, B. Pusey, J. Oi, A. M. Lask, L. P. Tomsho, M. D. Packard, F. Zhao, A. Sher, A. Tikhonov, B. Raney, N. Patterson, K. Linblad-Toh, E. S. Lander, J. R. Knight, G. P. Irzyk, K. M. Frederikson, T. T. Harkins, S. Sheridan, T. Pringle, and S. C. Schuster. 2008. Sequencing the Nuclear Genome of the Extinct Woolly Mammoth. *Nature* 456, 387–92.

Milner, A. R., and D. B. Norman. 1984. The Biogeography of Advanced Ornithopod Dinosaurs (Achosauria, Ornithischia)—A Cladistic-Vicariance Model. In W. E. Reif and F. Westphal, eds., *Third Symposium Mesozoic Terrestrial Ecosystems*, 146–51. Tubingen: Attempto.

Morell, V. 1993. Dino DNA: The Hunt and the Hype. *Science* 261, 160–62.

Mull, C. G., and K. E. Adams. 1989. *Bedrock Geology of the Eastern Koyukuk Basin, Central Brooks Range, and Eastcentral Arctic Slope along the Dalton Highway, Yukon River to Prudhoe Bay, Alaska*. Guidebook 7, vol. 2. Anchorage, AK: Division of Geological and Geophysical Surveys.

Mull, C. G., D. W. Houseknecht, and K. J. Bird. 2003. *Revised Cretaceous and Tertiary Stratigraphic Nomenclature in the Colville Basin, Northern Alaska*. U.S. Geological Survey Professional Paper 1673. http://pubs.usgs.gov/pp/p1673/p1673.pdf.

Nadon, G. C. 1993. The Association of Anastomosed Fluvial Deposits and Dinosaur Tracks, Eggs, and Nests: Implications for the Interpretation of Floodplain Environments and a Possible Survival Strategy for Ornithopods. *Palaios* 8, 31–44.

Nelms, L. G. 1989. Late Cretaceous Dinosaurs from the North Slope of Alaska. *Paleontology* 9, supplement 3, 34A.

Nessov, L. A. 1995. Dinosaurs of Northern Eurasia: New Data About Assemblages, Ecology, and Paleobiogeography. Trans. Tatayana Platonova. St. Petersburg, Russia: St. Petersburg State University.

Nikiforuk, A. 1984. Migration to a Watery Death. *Macleans* 97, 14–15.

Norman, D. B. 1985. *The Illustrated Encyclopedia of Dinosaurs*. London: Salamander.

Norman, D. B., H.-D. Sues, L. M. Witmer, and R. A. Coria. 2004. Basal Ornithopoda. In D. B. Weishampel, P. Dodson, and H. Osmólska, eds., *The Dinosauria*, 2nd ed., 393–412. Berkeley: University of California Press.

Norell, M. A., and P. J. Makovicky. 2004. Dromaeosauridae. In D. B. Weishampel, P. Dodson, and H. Osmólska, eds., *The Dinosauria*, 2nd ed., 196–209. Berkeley: University of California Press.

Norton, D. 2003. Environmental Ignorance: What It Costs. *Polar Times* 3, 9.

Núñez-Betelu, L. K., L. V. Hills, and P. J. Currie. 2005. The First Hadrosaur from the Late Cretaceous of Axel Heiberg Island, Canadian Arctic Archipelago. In D. R. Braman, F. Therrien, E. B. Koppelhaus, and W. Taylor, W., eds., *Short Papers, Abstracts, and Program, Special Publication: Dinosaur Park Symposium*, 69–72. Drumheller, Canada: Royal Tyrrell Museum.

Núñez-Betelu, L. K., L. V. Hills, F. F. Krause, and D. J. McIntyre. 1995. Upper Cretaceous Paleoshorelines of the Northeastern Sverdrup Basin, Ellesmere Island, Canadian Arctic Archipelago. In K. V. Smakov and D. K. Thurston, eds., *1994 Proceedings of the International Conference on Arctic Margins*,

43–49. Magadan, Russia: Russian Academy of Sciences.

Orth, D. J. 1967. *Dictionary of Alaskan Place Names.* U. S. Geological Survey Professional Paper 567. http://www2.borough.kenai.ak.us/AssemblyClerk/Assembly/Task%20Force/Flood%20Plain%20Task%20Force/042909%20Individual%20Items/Dictionary%20of%20Alaska%20Place%20Names%201.pdf.

Pääbo, S. 1993. Ancient DNA. *Scientific American* 269, 86–92.

Padian, K., and J. R. Horner. 2004. Dinosaur Physiology. In D. B. Weishampel, P. Dodson, and H. Osmólska, eds., *The Dinosauria,* 2nd ed., 660–71. Berkeley: University of California Press.

Parfitt, T. 2009. Russia's Polar Hero. *Science* 324, 1382–84.

Parrish, J. M., J. Totman Parrish, J. H. Hutchison, and R. A. Spicer. 1987. Late Cretaceous Vertebrate Fossils from the North Slope of Alaska and Implications for Dinosaur Ecology. *Palaios* 2, 377–89.

Pasch, A. D., and K. C. May. 2001. Taphonomy and Paleoenvironment of a Hadrosaur (Dinosauria) from the Matanuska Formation (Turonian) in South-Central Alaska. In D. H. Tanke and K. Carpenter, eds., *Mesozoic Vertebrate Life,* 219–36. Bloomington: Indiana University Press.

Paul, G. S. 1988. Physiological, Migratorial, Climatological, Geophysical, Survival, and Evolutionary Implications of Cretaceous Polar Dinosaurs. *Journal of Paleontology* 62, 640–52.

———. 1994. Physiology and Migration of North Slope Dinosaurs. In D. K. Thurston and K. Fujita, eds., *1992 Proceedings International Conference on Arctic Margins,* 405–408. Outer Continental Shelf-Mineral Management Service Publication 94-0040. Anchorage, AK: U. S. Department of the Interior.

———. 1997. Migration. In P. J. Currie and K. Padian, eds., *Encyclopedia of Dinosaurs,* 444–46. New York: Academic Press.

Phillips, R. L. 2003. Depositional Environments and Processes in Upper Cretaceous Nonmarine and Marine Sediments, Ocean Point Dinosaur Locality, North Slope, Alaska. *Cretaceous Research* 24, 499–523.

Poinar, G. O. 2005. A Cretaceous Palm Bruchid, *Esopachymerus antiqua,* n. gen., n. s (Coleoptera: Bruchidae: Pachymerini) and Biogeographical Implications. *Proceedings of the Entomological Society of Washington* 107, 392–97.

Poinar, G. O., and R. Poinar. 1994. *The Quest for Life in Amber.* Reading, MA: Addison-Wesley.

Prasad, V., C. A. E. Strömberg, A. Habib, and A. Sahni. 2005. Dinosaur Coprolites and the Evolution of Grasses and Grazers. *Science* 310, 1177–80.

Ralrick, P. E. and D. H. Tanke. 2008. Comments on the Quarry Map and Preliminary Taphonomic Observations of the *Pachyrhinosaurus* (Dinosauria: Ceratopsidae) Bone Bed at Pipstone Creek, Alberta, Canada. In P. J. Currie, W. Langston Jr., and D. H. Tanke, eds., *A New Horned Dinosaur From an Upper Cretaceous Bone Bed in Alberta,* 109–16. Ottawa, Canada: NRC Research Press.

Raup, D. M. 1991. *Extinction: Bad Genes or Bad Luck?* New York: Norton.

Rich, T. H. 2008. Tunnelling for Dinosaurs in the High Arctic. *Deposits Magazine,* January, 18–22.

Rich, T. H., R. A. Gangloff, and W. R. Hammer. 1997. Polar Dinosaurs. In P. J. Currie and K. Padian, eds., *Encyclopedia of Dinosaurs,* 562–73. San Diego, CA: Academic Press.

Rich, T. H., and P. Vickers-Rich. 2000. *Dinosaurs of Darkness.* Bloomington: Indiana University Press.

Rich, T. H., P. Vickers-Rich, and R. A. Gangloff. 2002. Science's Compass, Polar Dinosaurs. *Science* 295, 979–80.

Ridgway, K. D., J. M. Trop, and A. R. Sweet. 1997. Thrust-Top Basin Formation along a Suture Zone, Cantwell Basin, Alaska Range: Implications for the Development of the Denali Fault System. *Geological Society of America Bulletin* 109, 505–23.

Roderick, J. 1997. *Crude Dreams: A Personal History of Oil and Politics in Alaska.* Seattle, WA: Epicenter Press.

Roehler, H. W., and G. D. Stricker. 1984. Dinosaur and Wood Fossils from the Cretaceous Corwin Formation in the National Petroleum Reserve, North Slope, Alaska. *Journal of the Alaska Geological Society* 4, 35–41.

Rogers, R. R., and S. M. Kidwell. 2007. A Conceptual Framework for the Genesis and Analysis of Vertebrate Skeletal Concentrations. In R. R. Rogers, D. A. Eberth, and A. R. Fiorillo, eds., *Bonebeds: Genesis, Analysis, and Paleobiological Significance,* 1–64. Chicago: University of Chicago Press.

Rozhdestvenskiy, A. K. 1973. The Study of Cretaceous Reptiles in Russia. *Paleontology Journal* 2, 206–14.

Rudwick, M. J. S. 1985. *The Meaning of Fossils: Episodes in the History of Palaeontology.* 2nd ed. Chicago: University of Chicago Press.

Russell, D. A. 1984. A Check List of the Families and Genera of North American Dinosaurs. *Syllogeus* 53. Ottawa: National Museums of Canada.

Russell, D. A., and J. F. Bonaparte. 1997. Dinosaurian Faunas of the Later Mesozoic. In J. O. Farlow and M. K. Brett-Surman. *The Complete Dinosaur*, 645–60. Bloomington: Indiana University Press.

Russell, D. A., and P. Dodson. 1997. The Extinction of the Dinosaurs: A Dialogue between a Catastrophist and a Gradualist. In J. O. Farlow and M. K. Brett-Surman. *The Complete Dinosaur*, 662–72. Bloomington: Indiana University Press.

Ryan, M. J., P. J. Currie, J. D. Gardner, M. K. Vickaryous, and J. M. Lavigne. 1998. Baby Hadrosaurid Material Associated with an Unusually High Abundance of *Troodon* Teeth from the Horseshoe Canyon Formation, Upper Cretaceous, Alberta, Canada. *GAIA* 15, 123–33.

Ryan, M. J., and A. Russell. 2001. Dinosaurs of Alberta (Exclusive of Aves). In D. H. Tanke, K. Carpenter, and M. W. Skrepnick, eds., *Mesozoic Vertebrate Life*. Bloomington: Indiana University Press, 279–97.

Rylaarsdam, J. R., B. L.Varban, A. G. Plint, L. G. Buckley, and R. T. McCrea. 2006. Middle Turonian Dinosaur Paleoenvironments in the Upper Kaskapau Formation, Northeast British Columbia. *Canadian Journal of Earth Sciences* 43, 631–52.

Sagan, C. 1996. *The Demon-Haunted World: Science as a Candle in the Dark*. New York: Ballantine.

Sampson, S. D., and M. A. Loewen. 2010. Unraveling a Radiation: A Review of the Diversity, Stratigraphic Distribution, Biogeography, and Evolution of Horned Dinosaurs (Ornithischia: Ceratopsidae). In M. J Ryan, B. J. Chinnery-Allgeier, and D. A. Eberth, eds., *New Perspectives on Horned Dinosaurs: The Royal Tyrrell Museum Ceratopsian Symposium*, 405–27. Bloomington: Indiana University Press.

Sanborn, W. B. 1976. *Oddities of the Mineral World*. New York: Van Nostrand Reinhold.

Sarjeant, W. A. S. 1997. The Earliest Discoveries. In J. O. Farlow and M. K. Brett-Surman, eds., *The Complete Dinosaur*, 9–11. Bloomington: Indiana University Press.

Schweitzer, M. H. 1997. Molecular Paleontology: Rationale and Techniques for the Study of Ancient Biomolecules. In J. O. Farlow, and M. K. Brett-Surman, eds., *The Complete Dinosaur*, 136–49. Indiana University Press.

Schweitzer, M. H., C. Johnson, T. G. Zocco, J. R. Horner, and J. R. Starkey. 1997. Preservation of Biomolecules in Cancellous Bone of *Tyrannosaurus rex*. *Journal of Vertebrate Paleontology* 17, 349–59.

Scotese, C. R., L. M. Gahagan, M. I. Ross, J.-Y. Royer, R. D. Müeller, D. Nürnberg, C. L. Mayes, L. A. Lawver, R. L. Tomlins, J. S. Newman, C. E. Heubeck, J. K. Winn, L. Beckley, and J. G. Sclater. 1987. *Atlas of Mesozoic and Cenozoic Plate Tectonic Reconstructions*. Austin: Technical Report by University of Texas at Austin, Institute of Geophysics, Paleogeographic Mapping Project.

Senkowsky, S. 2003. Arctic Dinosaurs—Inauguration to Field Work. *Polar Times* 3, 6–7.

Shinkarev, L. 1973. *The Land Beyond the Mountains: Siberia and Its People Today*. New York: Macmillan.

Smith, C. 1993. *The Alaska Almanac: Facts About Alaska* 17th ed. Anchorage AK: Northwest Books.

Smith, A. G., A. M. Hurley, and J. C. Briden. 1981. *Phanerozoic Paleocontinental World Maps*. London: Cambridge University Press.

Sorensen, B. 2000. Unearthing Secrets. *Winds of Change* 15, 18–21.

Spicer, R. A. 2003. A Greenhouse Case Study. In P. W. Skelton, R. A. Spicer, S. P. Kelley, and I. Gilmour, eds., *Evolving Life and the Earth*, 128–62. Cambridge: Cambridge University Press.

Spicer, R. A., and A. B. Herman. 2010. The Late Cretaceous Environment of the Arctic: A Quantitative Reassessment Based on Plant Fossils. *Palaeogeography, Palaeoclimatology, Palaeoecology* 295, 423–42.

Spicer, R. A., and J. Totman Parrish. 1990. Late Cretaceous-Early Tertiary Palaeoclimates of Northern High Latitudes: A Quantitative Review. *Geological Society of London Journal* 147, 329–41.

Spicer, R. A., J. Totman Parrish, and P. R. Grant. 1992. Evolution of Vegetation and Coal-Forming Environments in the Late Cretaceous of the North Slope of Alaska. In P. J. McCabe, and J. Totman Parrish, eds., *Controls on the Distribution and Quality of Cretaceous Coals*, 177–86. Geological Society of America Special Paper 267. Boulder, CO: Geological Society of America.

Stott, D. F. 1975. *The Cretaceous System in North-Eastern British Columbia*. Geological Association of Canada Special Paper 13, 441–67. Waterloo, Canada: Geological Association of Canada.

———. 1984. Cretaceous Sequences of the Foothills of the Canadian Rocky Mountains. In D. F. Stott and D. J. Glass, eds., *The Mesozoic of Middle North America*, 85–107. Canadian Society of Petroleum Geologists, Memoir 9.

Strahler, A. H., and A. N. Strahler. 1991. *Modern Physical Geography*. 4th ed. New York: John Wiley.

Suslov, S. P. 1961. *Physical Geography of Asiatic Russia*. San Francisco: W. H. Freeman.

Tanke, D. H. 2007. Ceratopsian Discoveries and Work in Alberta, Canada. In D. B. Brinkman, M. J. Ryan, B. Chinnery-Allgeier, D. A. Eberth, and P. J. Currie, eds., *Ceratopsian Symposium*. Drumheller, Canada: Royal Tyrrell Museum of Palaeontology.

Tiffney, B. H. 1997. Land Plants as Food and Habitat in the Age of Dinosaurs. In J. O. Farlow and M. K. Brett-Surman, eds., *The Complete Dinosaur*, 352–70. Bloomington: Indiana University Press.

Totman Parrish, J. 1998. *Interpreting Pre-Quaternary Climate from the Geologic Record*. New York: Columbia University Press.

Totman Parrish, J., and R. A. Spicer. 1988. Late Cretaceous Terrestrial Vegetation: A Near-Polar Temperature Curve. *Geology* 16, 22–25.

Townsend, J. H., J. Slapcinsky, L. Krysko, E. M. Donlan, and E. A. Golden. 2005. Predation of a Tree Snail *Drymaeus multilineatus* (Gastropoda: Bulimulidae) by *Iguana iguana* (Reptilia: Iguanidae) on Key Biscayne, Florida. *Southeastern Naturalist* 4, 361–64.

Troyer, K. 1984. Diet Selection and Digestion in *Iguana iguana*: The Importance of Age and Nutrient Requirements. *Oecologia* 61, 201–207.

Turco, R. P., O. B. Toon, T. P. Ackerman, J. B. Pollack, and C. Sagan. 1983. Nuclear Winter: Global Consequences of Multiple Nuclear Explosions. *Science* 222, 1283–92.

Upchurch, G. R., B. L. Otto-Bliesner, and C. R. Scotese. 1999. Terrestrial Vegetation and its Effects on Climate during the Latest Cretaceous. In E. Barrera and C. C. Johnson, eds., *Evolution of the Cretaceous Ocean-Climate System*, 407–26. Geological Society of America Special Paper 332. Boulder, CO: Geological Society of America.

U. S. Census Bureau. 2010. http://quickfacts.census.gov/qfd/states/02/02185.html.

Valkenburg, P. 1998. Herd Size, Distribution, Harvest, Management Issues, and Research Priorities Relevant to Caribou Herds in Alaska. *Rangifer* 10, 125–29.

Van Andel, T. H. 1985. *New Views on an Old Planet: Continental Drift and the History of the Earth*. New York: Cambridge University Press.

Vandermark, D., J. A. Tarduno, D. B. Brinkman, R. D. Cottrell, and S. Mason. 2009. New Late Cretaceous Macrobaenid Turtle with Asian Affinities from the High Canadian Arctic: Dispersal Via Ice-free Polar Routes. *Geology* 37, 183–86.

Van Itterbeeck, J., Y. Bolotsky, P. Bultynck, and P. Godefroit, P. 2005. Stratigraphy, Sedimentology and Palaeoecology of the Dinosaur-Bearing Kundur Section (Zeya-Bureya Basin, Amur Region, Far Eastern Russia. *Geological Magazine* 142, 735–50.

Varricchio, D. J. 1997. Troodontidae. In P. J. Currie and K. Padian, eds., *Encyclopedia of Dinosaurs*, 749–54. San Diego, CA: Academic Press.

Varricchio, D. J., and J. R. Horner. 1993. Hadrosaurid and Lambeosaurid Bone Beds from the Upper Cretaceous Two Medicine Formation of Montana: Taphonomic and Biologic Implications. *Canadian Journal of Earth Sciences* 30, 997–1006.

Varricchio, D. J., F. Jackson, B. Scherzeer, and J. Shelton. 2005. Don't Have a Cow, Man! It's Only Actualistic Taphonomy on the Yellowstone River of Montana. *Journal of Vertebrate Paleontology* 25, supplement 3, 126A.

Varricchio, D. J., A. J. Martin, and Y. Katsura. 2007. First Trace and Body Fossil Evidence of a Burrowing, Denning Dinosaur. *Proceedings of the Royal Society of London* 274B, 1361–68.

Velikovsky, I. 1955. *Earth in Upheaval*. New York: Dell.

Vickaryous, M. K., T. Maryańska, and D. B. Weishampel. 2004. Ankylosauria. In D. B. Weishampel, P. Dodson, and H. Osmólska. 2004. *The Dinosauria*, 2nd ed., 363–92. Berkeley: University of California Press.

Wahrhaftig, C. 1965. *Physiographic Divisions of Alaska*. U. S. Geological Survey Professional Paper 482.

Ward, P. D. 1992. *On Methuselah's Trail*. New York: W. H. Freeman.

Weigelt, J. 1989. *Recent Vertebrate Carcasses and Their Paleobiological Implications*. Chicago: University of Chicago Press.

Weishampel, D. B. Dinosaurian Distribution. 1990. In *The Dinosauria*, 63–139. Berkeley: University of California Press.

Weishampel, D. B., P. M. Barrett, R. A. Coria, J. Le Loeuff, X. Xing, Z. Xijin, A. Sahni, E. M. P. Gomani, C. R. Noto. 2004. Dinosaur Distribution. In *The Dinosauria*, 2nd ed., 517–683. Berkeley: University of California Press.

Weishampel, D. B., P. Dodson, and H. Osmólska. 1990. *The Dinosauria*. Berkeley: University of California Press.

———. 2004. *The Dinosauria*. 2nd ed. Berkeley: University of California Press.

Williams, G. 2003. *Voyages of Delusion: The Quest for the Northwest Passage*. New Haven, CT: Yale University Press.

Williams, H. B., and C. R. Stelck. 1975. Speculations on the Cretaceous Palaeogeography of North America. In W. G. E. Caldwell, ed., *The Cretaceous System in the Western Interior of North America*, 1–20. Geological Association of Canada Special Paper 13. Waterloo, Canada: Geological Association of Canada.

Williams, V. S., P. M. Barrett, and M. A. Purnell. 2009. Quantitative Analysis of Dental Microwear in Hadrosaurid Dinosaurs, and the Implications for Hypotheses of Jaw Mechanics and Feeding. *Proceedings of the National Academy of Sciences* 106, 11194–99.

Witmer, L. M., and R. C. Ridgely. 2008. Structure of the Brain Cavity and Inner Ear of the Centrosaurine Ceratopsid Dinosaur *Pachyrhinosaurus* Based on CT Scanning and 3D Visualization. In P. J. Currie, W. Langston Jr., and D. H. Tanke, eds., *A New Horned Dinosaur from an Upper Cretaceous Bone Bed in Alberta*, 117–44. Ottawa, Canada: NRC Research Press.

Witte, K. W., D. B. Stone, and C. G. Mull. 1987. Paleomagnetism, Paleobotany, and Paleogeography of the Cretaceous, North Slope, Alaska. In vol. 1 of I. Tailleur and P. Weimer, eds., *Alaska North Slope Geology*, 571–79. Bakersfield, CA: Society of Economic Paleontologists and Mineralogists and the Alaska Geological Society.

Woodward, S. R., N. J. Weyland, and M. Bunnell. 1994. DNA Sequence from Cretaceous Period Bone Fragments. *Science* 266, 1226–32.

Wyllie, P. J. 1976. *The Way the Earth Works: An Introduction to the New Global Geology and its Revolutionary Development*. New York: John Wiley.

Xiao-chun, W., and H.-D Sues. 1996. Anatomy and Phylogenetic Relationship of *Chimaerasuchus paradoxus*, an Unusual Crocodyliform Reptile from the Lower Cretaceous of Hubei, China. *Journal of Vertebrate Paleontology* 16, 688–702.

Yang, H. 1997. Ancient DNA from Pleistocene Fossils: Preservation, Recovery, and Utility of Ancient Genetic Information for Quaternary Research. *Quaternary Science Reviews* 16, 1145–61.

Yorath, C. J., and D. G. Cook. 1981. *Cretaceous and Tertiary Stratigraphy and Paleogeography, Northern Interior Plains, District of Mackenzie*. Geological Survey of Canada Memoir 398.

INDEX

ROLAND A. GANGLOFF retired in 2003 as associate professor from the Department of Geology and Geophysics and as curator of earth science at the University of Alaska Museum of the North in Fairbanks. He is presently a visiting scholar at the University of California Museum of Paleontology in Berkeley, where he continues to indulge his two research passions, Early Cambrian reef-forming organisms and Arctic dinosaurs. Gangloff has engaged in fieldwork throughout Alaska, southern Yukon Territory, Alberta, Canada, and Siberia. He has published numerous technical and popular articles on the dinosaurs of Alaska.